MATTER, LIFE, AND GENERATION

MATTER, LIFE, AND GENERATION

Eighteenth-century embryology and the Haller–Wolff debate

SHIRLEY A. ROE

Department of the History of Science
Harvard University

CAMBRIDGE UNIVERSITY PRESS

Cambridge
London New York New Rochelle
Melbourne Sydney

PUBLISHED BY THE PRESS SYNDICATE OF THE UNIVERSITY OF CAMBRIDGE
The Pitt Building, Trumpington Street, Cambridge, United Kingdom

CAMBRIDGE UNIVERSITY PRESS
The Edinburgh Building, Cambridge CB2 2RU, UK
40 West 20th Street, New York NY 10011–4211, USA
477 Williamstown Road, Port Melbourne, VIC 3207, Australia
Ruiz de Alarcón 13, 28014 Madrid, Spain
Dock House, The Waterfront, Cape Town 8001, South Africa

http://www.cambridge.org

First published 1981
First paperback edition 2002

A catalogue record for this book is available from the British Library

Library of Congress Cataloguing in Publication data
Roe, Shirley A. 1949–
Matter, life, and generation.
Bibliography: p.
Includes index.
1. Embryology – History. 2. Biology –
Philosophy. 3. Haller, Albrecht von, 1708–1777.
4. Wolff, Caspar Friedrich, 1733–1794. I. Title.
QL953.R64 591.3′3′09 80-19611

ISBN 0 521 23540 5 hardback
ISBN 0 521 52525 X paperback

Contents

Illustrations

Preface

This study grew out of a desire to examine a philosophical problem, the relation between theory and observation in science, within a historical context. Both historians and philosophers have become increasingly cognizant in recent years of the influence that philosophical, religious, social, and other nonobservational issues exert in the practice of science. Yet much work remains to be done in examining the specific paths such influences have taken in the history of science.

This book focuses on the Haller–Wolff debate over embryological development as a case study in scientific controversy. Haller and Wolff, as representatives of two eighteenth-century philosophical traditions, brought to their embryological controversy a whole host of "extrascientific" assumptions and expectations, which fundamentally colored the observational level of their debate. Ultimately, what was at stake in their confrontation was the nature of scientific explanation and how it should be applied to the phenomena of embryology. My treatment of the Haller–Wolff debate analyzes these various levels of controversy in an attempt to assess the relative roles they played in the course and outcome of the debate itself. I argue, moreover, that the Haller–Wolff debate is not an isolated event, but that philosophical questions have repeatedly played a guiding role in embryological thought. Thus, I see my study as a contribution not only to the history of embryology, but to Enlightenment intellectual history more generally.

I owe a very special debt of gratitude to the late Erich Hintzsche, who introduced me to the Haller materials in Bern during my visits there in 1973 and 1974, and whose dedication to Haller studies has served as an inspiration to all who have worked on this great figure in recent years. I would also like to thank H. A. Haeberli of the Burgerbibliothek in Bern and P. Monnier of the Bibliothèque Publique et Universitaire in Geneva for their kind assistance during my research in these libraries. I am indebted to Tat'jana A. Lukina for providing me with a copy of her published edition of one of Wolff's manuscript treatises, without which my study would have remained incomplete. Finally, I would like to thank Richard J. Wolfe and the staff of the Rare Books Room at the Francis A. Countway Library of Medicine, Harvard Medical School, for making available to me their superb collection of eighteenth-century sources.

Donald F. McCabe of the Harvard Classics Department was of special help in translating portions of Wolff's unpublished materials and Wolff's letters to Haller (see Appendix B of this book). I would like to thank Patricia L. Chaput of the Harvard Slavic Department for guiding me through Gaissinovitch's Russian-language book on Wolff, and Charles W. Qwatt for his assistance with regard to some of my German translations. Appreciation is gratefully extended to the Josiah Macy, Jr. Foundation and to the Harvard Graduate Society for fellowship and research aid at various points in this project. Finally, let me thank several individuals for their encouragement and intellectual inspiration during my years of research and writing, among them, Frederick Churchill, I. Bernard Cohen, Renato G. Mazzolini, Everett Mendelsohn, Jane M. Oppenheimer, Dov Ospovat, Barbara Gutmann Rosenkrantz, and Frank J. Sulloway. Permission is gratefully acknowledged from D. Reidel Publishing Company for the use of material previously published in the *Journal of the History of Biology* (1975, 1979); and from the Burgerbibliothek in Bern, the Bibliothèque Publique et Universitaire in Geneva, and the archives of the Germanisches Nationalmuseum in Nüremberg for permission to quote from unpublished manuscript sources.

S. A. R.

1

Introduction: mechanism and embryology

In 1683, the French savant Bernard de Fontenelle offered the following rejoinder to the mechanical physiology of his day: "Do you say that beasts are machines just as watches are? Put a male dog-machine and a female dog-machine side by side, and eventually a third little machine will be the result, whereas two watches will lie side by side all their lives without ever producing a third watch" (1683:312). The ability of living organisms to recreate themselves is perhaps the most striking and most distinguishing feature of life. Not surprisingly, explaining the phenomenon of reproduction has remained a central issue in biology throughout the ages.

Fontenelle's comments were made just four years before the publication of Newton's *Principia*, at the height of the Scientific Revolution. The success of mechanical explanations in the physical and astronomical sciences led some investigators to apply this kind of reasoning to biology as well. Could life phenomena be explained on the basis of matter and motion? Were living organisms simply highly organized machines? Locomotion, sensation, the circulation of blood, the movement of fluids in plants, digestion, respiration – all were subjected to mechanical analysis. Reproductive phenomena were no exception, although the application of mechanism proved more problematic here.

In the late seventeenth and early eighteenth centuries, two rival schools of thought on the subject of generation existed. The preformationists believed that the embryo preexists in some form in either the maternal egg or the male spermatozoon.[1] Most also thought that all embryos had been formed by God at the Creation and encased within one another to await their future appointed time of development. Epigenesists, on the other hand, argued that each embryo is newly produced through gradual development from unorganized material. Various explanations were proposed for how this gradual formation is accomplished, yet epigenesists were united in their opposition to preexistence.

Although clashes occurred between preformationists and epigenesists throughout the Enlightenment, perhaps the most important of these was the debate that took place between Albrecht von Haller (1708–77) and Caspar Friedrich Wolff (1734–94). Haller, a renowned scientific figure, announced his support for preformation in 1758, just one year prior to the publication of Wolff's doctoral dissertation, which strongly endorsed epigenesis. The ensuing controversy, lasting for over a decade, crystallized many of the key issues of eighteenth-century embryology. The role of mechanism in biological explanation, the relationship of God to his Creation, the question of spontaneous generation, and the problems of regeneration, hybrids, and monstrous births – all these were points of issue in the Haller–Wolff debate.

More importantly, the controversy between Haller and Wolff illustrates the fundamental tie between biological and philosophical questions that existed in the Enlightenment period. Philosophical concerns were in fact largely responsible for the rise and popularity of preformationist theories over epigenesis. The clash between Haller and Wolff epitomizes this philosophical nature of eighteenth-century embryology, for Haller and Wolff came from widely divergent philosophical backgrounds. Haller, a Newtonian mechanist and a deeply religious man, held beliefs about the nature of the world and about scientific explanation that differed fundamentally from those of Wolff, whose own viewpoint derived largely from the tradition of German rationalism. Their debate over embryological development can be fully understood only when viewed as a controversy over these underlying philosophical differences. Furthermore, as representatives of two major Enlightenment schools of thought, Haller and Wolff illustrate two important cases of the ways in which philosophical issues guided much of eighteenth-century embryology. As such, an analysis of their work and their controversy sheds light on a number of aspects of biological thought during this period.

THE RISE OF PREFORMATION THEORIES

Although there were those who, before the late seventeenth century, believed that the embryo was in some fashion preformed in the body of the parent before conception, the notion that all embryos had existed from the beginning of the world was first formulated in the 1670s with the work of Malebranche, Swammerdam, Perrault, and others. These theories of preexistence, based for the most part on the concept of *emboîtement* (encasement), did not grow directly out of the earlier preformationist positions (see note 1). Rather, they arose in response to a set of difficulties and concerns that were prompted by the appearance in the mid-seventeenth century of several epigenetic theories of development, propounded by Harvey, Descartes, Highmore, Borelli, and others.

Both Aristotle and Galen had proposed theories of epigenesis, albeit with important differences between their two systems[2]; and it is largely within the context of Renaissance scholasticism, the heir to these two great thinkers, that William Harvey's *Exercitationes de generatione animalium* (1651) must be understood. Written largely as a commentary on the theories of Aristotle and Fabricius ab Aquapendente, Harvey's work was based on observations that he had made on deer and on incubated chicken eggs. Harvey combated the Galenic and Hippocratic two-semen theory then in vogue by claiming that he could find no female semen and by showing that in female deer dissected shortly after copulation there was no evidence of male semen entering the uterus. Consequently, the embryo could not be the result of the mixing together of male and female seminal material. Furthermore, through his famous dictum "Omne vivum ex ovo," Harvey proposed that all organisms, viviparous as well as oviparous, develop from a primordial egg. Not a preexisting germ, the egg was thought by Harvey to be a product of conception. Finally, Harvey's detailed observations of day-by-day development of chick embryos led him to conclude "that the generation of the chick from the egg is the result of epigenesis . . . and that all its parts are not fashioned simultaneously, but emerge in their due succession and order" (1651, 1847 trans.: 336).[3]

Although Harvey's work provided significant improvements over the theories of his predecessors, his views had limited success, partly because, by 1651, his Aristotelian mode of argumentation had begun to seem out of date. Contemporaneous with Harvey, however, was another promulgator of epigenetic development, René Descartes, who was the first to offer an explanation for generation based solely on matter and motion. Descartes dealt with animal physiology in his *Traité de l'homme*, the second part of his *Monde*, which was written in the 1630s but suppressed by Descartes because of its Copernican viewpoint. (Galileo had recently been condemned for like views by the Catholic church.) Published after his death, Descartes's *Traité de l'homme* proposed mechanical explanations for the phenomena of sensation, muscular movement, digestion, the circulation of the blood, and other vital functions.[4] Yet generation was left untreated, for Descartes remarked in a letter to Mersenne that he had given up trying to deal with this subject in his treatise (Descartes 1964–74, 1:254). In the late 1640s, when Descartes returned again to generation, he described this earlier stage in his thinking about the animal organism, noting "I had almost lost hope of finding the causes of its formation." Yet, he proclaimed in this same letter, "in meditating thereupon, I have discovered so much new land, that I hardly doubt that I can complete the whole of the Physics according to my desire" (5:261). Descartes saw the generation of animals as the last segment to come under the wing of his mechanical philosophy, forming there the completion of Cartesian physiology.

The result of Descartes's deliberations in the late 1640s was a small treatise *De la formation de l'animal*, published posthumously along with the *Traité de l'homme* in 1664. Here Descartes proposed an explanation for development based entirely upon the movement of particles. According to Descartes's system, reproduction begins with the mixing of semen from the male and female, resulting in a fermentation of particles. The movement of these particles leads to the formation of the heart, followed by the other embryonic parts. "If one knew what all the parts of the semen of a certain species of animal are, in particular, for example, of man," Descartes declared, "one could deduce from this alone, by reasons entirely mathematical and certain, the whole figure and conformation of each of its members" (1664: 146). Through matter and motion alone, one can explain not only the inanimate but the animate world as well. "It is no less natural for a clock, made of a certain number of wheels, to indicate the hours," he proclaimed, "than for a tree born from a certain seed, to produce a particular fruit" (1644:326).

Descartes's explanation for embryological development by mechanical causation was not a successful one. In particular, it failed, as did those of other mechanistic epigenesists, to explain *why* development proceeds as it does, that is, why the proper organism, with its parts perfectly arranged, is formed. How does a process based on matter and motion alone result in a complex living organism?

This insufficiency of mechanical explanations of gradual development was an important element in the rise of preexistence theories. As Nicolas Malebranche noted in 1688 concerning Descartes's theory of epigenesis, "The rough sketch given by this philosopher may help us understand how the laws of motion are sufficient to bring about the gradual growth of the parts of the animal. But that these laws could form them and link them together is something that no one will ever prove" (1688:264). It does not seem possible that mechanical laws could both fashion and organize the parts of the organism. "It is easy to see," Malebranche declared, "that the general laws of the communication of motion are too simple for the construction of organic bodies" (p. 252). Mechanical causes may be part of the process of development, but they cannot account for reproduction itself.

Malebranche was the first to fully articulate, in 1674, a theory of preformation by *emboîtement*. In discussing the limits of our senses, especially vision, Malebranche turned to biological examples. If one looks closely at a tulip bulb, using a magnifying lens, one can see all the parts of the future tulip folded up in miniature inside the bulb. And one can assume that the same may be the case in the seeds of all trees and plants. "It does not even seem unreasonable," Malebranche declared, "to think that there are infinite trees in a single germ, since it does not contain only the tree which is the seed, but also a very great number of other seeds, which are all enclosed in those of the new tree. . . . one can say that in a single apple pit, there would be apple trees, apples, and the seeds of apples for infinite or almost infinite centuries" (1674:82). Each seed, then, would contain the seeds of all future individuals, encased within one another. This notion can be extended to cover animals as well: "One sees in the germ of the bulb of a tulip the entire tulip. One sees also in the germ of a fresh egg, and which has not been covered, a chicken which is perhaps entirely formed. One sees frogs in the eggs of frogs, and one will see other animals in their germs, when one has appealed to and experimented enough to discover them" (pp. 82-83). From all of this evidence we can conclude, according to Malebranche, "that all the bodies of men and of animals, which have been born up to the consummation of the century, have perhaps been produced as long ago as the creation of the world" (p. 83; see also Schrecker 1938).

In his discussion of animal *emboîtement*, Malebranche referred to the work of Marcello Malpighi on chick eggs and of Jan

Swammerdam on frogs. Malpighi had shown in his *Dissertatio epistolica de formatione pulli in ovo* (1673) that one can observe, in eggs that have been fertilized but not yet incubated, the rudiments of the embryo already formed. Although Malpighi never argued for the preexistence of embryos before fertilization (and in fact reported that one can see nothing in the unfertilized egg), his observations were quickly taken up, by Malebranche and others, as evidence for preformation. By the eighteenth century, Malpighi was regularly cited as a preformationist. That the rudiments of the chick embryo could be seen from the earliest moments became a principal example in support of the preexistence of all germs.[5]

Even more important, Malpighi contributed the first detailed series of observations on the development of chick embryos.[6] Before Malpighi, Aldrovandi, Coiter, Fabricius ab Aquapendente, Harvey, and others had based their work on observational investigations; yet Malpighi's careful hour-by-hour descriptions of developing embryos, accompanied by superb illustrations (see Figure 1), represented a major advance over prior attempts to understand embryological development. Many of Malpighi's drawings are so clear and so detailed that they stand up well even when compared with modern-day diagrams. Observation and description were Malpighi's forte, and his declining to elaborate a clear theoretical position in his treatises undoubtedly contributed to later confusion over his views on preformation.

In a similar way, Swammerdam's observations on frogs' eggs and on the development of insects were seen as providing important evidence for preformation. In his book on insects published in 1669, Swammerdam claimed that one could observe the structure of the butterfly folded up within the chrysalis, and the wings and other parts in the dissected caterpillar. Initially directed against the theory of metamorphosis advocated by Harvey and others (that the caterpillar becomes transformed all at once into the butterfly), Swammerdam's observations were used by preformationists as an example of development from preexisting parts. Similarly, Swammerdam's brief comments on frogs' eggs made in 1672, where he remarked that the black spot in the egg is "the frog itself complete in all its parts" (1672:21), were cited by Malebranche and others as further evidence in support of preexistence.

Although Swammerdam expressed support for the notion of *emboîtement* in brief passages in 1669 and 1672 (see Roger 1963:

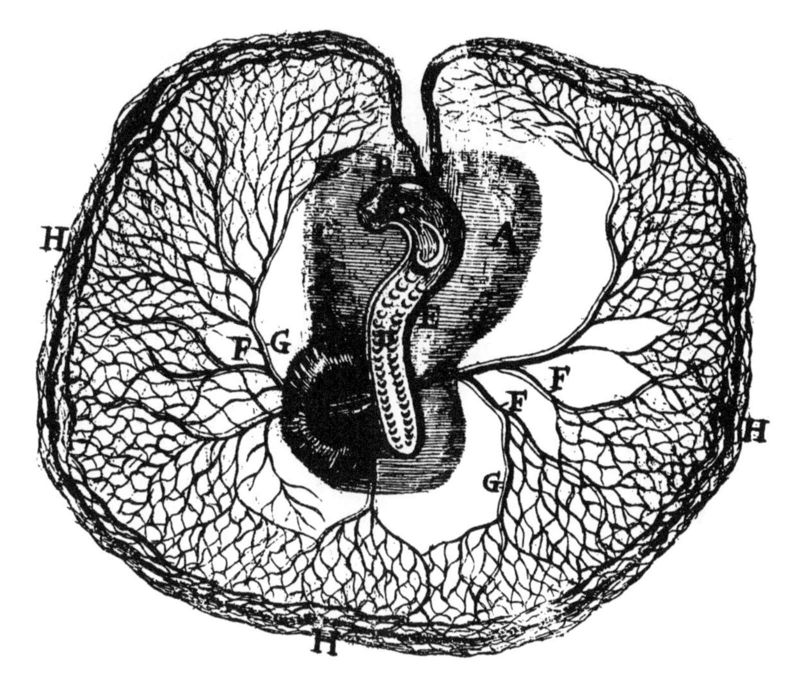

Figure 1. Malpighi's illustration of a chick embryo at 62 hours, showing the blood vessel network extending out over the yolk in the area vasculosa. (From *Opera omnia*, 1686; reproduced by permission of the Houghton Library, Harvard University)

334–35; Bowler 1971:238), his own principal concern was to combat the theories of insect metamorphosis and spontaneous generation. Yet the fact that his and Malpighi's observations were immediately taken up by those making preformationist claims indicates that the concept of preexistence was not one that grew out of observational evidence alone. It is clear, as several scholars have pointed out, that preformation through preexistence was a theory that responded more to philosophical than to observational needs.[7] As in the case of Malebranche, one of the principal motivations behind preformationist theories was the need to combat the implications that a fully mechanistic epigenetic position entailed. Although it was widely believed in the late seventeenth century that the universe must operate through mechanical laws, it was also felt that these laws were not sufficient to account for its origins and especially for the construction of living organisms. Claude Perrault, another early proponent of development through preexisting germs (although in his case through panspermism rather than encasement), made explicit his concern over the limits of mechanical laws. As he remarked in 1680, "I do not know if one can comprehend how a work of this quality would be the effect of the ordinary forces of nature. . . . for I find finally that it is scarcely more inconceivable . . . that the world has been able to form itself from matter out of chaos, than an ant can form another from the homogeneous substance of the semen from which it is believed to be engendered" (p. 481). For Perrault, a mechanical universe could not possibly be responsible on its own for the formation of living creatures. George Garden, an early supporter of animalculist preformation, claimed in a similar vein that "all the laws of Motion which are as yet discovered, can give but a very lame account of the forming of a Plant or Animal. We see how wretchedly Des Cartes came off when he began to apply them to this Subject; they are form'd by Laws yet unknown to Mankind, and it seems most probable that the *Stamina* of all the Plants and Animals that have been, or ever shall be in the World, have been formed *ab Origine Mundi* by the Almighty Creator within the first of each respective kind" (1691:476–77).

The rise of preformationist theories in the late seventeenth century was thus a response principally to a series of philosophical problems posed by the application of mechanical explanation to embryology. It was not only that mechanical epigenesis seemed incapable of accounting for why the embryo develops as it does, although this was an important problem. But even more significantly, it was the implications of a fully mechanistic embryology that were the most disturbing. If matter could form organized beings, living creatures, then what role was left for the Divine Creator? Preformation through *emboîtement* provided a solution to this difficulty while still preserving a mechanical universe. All organisms were formed by God at the Creation and encased within one another, so that at the appointed time each tiny preformed embryo could expand and develop, through mechanical means, into a full-fledged organism. Preexistence avoided the atheistic and materialistic implications of development by epigenesis, while also accounting for the source of animal organization. Embryos develop into the proper organisms because all of their parts were created at one time and arranged in the proper fashion by God.

The impossibility of mechanistic epigenesis was further enhanced by the identification, commonly made, between mechanism and blind chance. Matter was viewed by most as entirely passive, put into motion only through mechanical laws. But since these laws of motion are blind, that they could know how to form a living organism seemed out of the question. Both self-active matter and a God actively involved in each instance of generation were ruled out in the mechanistic universe of late-seventeenth-century thinkers. The theory of preformation offered the only account of embryological development consistent with this view of a divinely created, mechanically operating world.

By the beginning of the eighteenth century, the theory of preformation was widely accepted. Animalculism was adopted by some, notably Hartsoeker, Nicolas Andry, Boerhaave, and Leibniz; yet the notion of preexistent germs in the male spermatozoa never made the same headway that ovist preformation did. The principal difficulty lay in the tremendous waste that male *emboîtement* would imply. Why would God have created so many creatures in spermatozoa destined never to develop? Regnier de Graaf's work on the female reproductive organs, published in 1672, had established that mammals develop, like oviparous animals, from a female egg, although the actual mammalian ovum was not observed until the nineteenth century.[8] Ovist preformation theories were widely adopted in the early eighteenth century, prompting both new observational studies on egg development and continued support for *emboîtement*.

REGENERATION AND THE FRESHWATER POLYP

During the first half of the eighteenth century, regeneration became a subject of increasing study. Both Thévenot and Perrault had described regeneration in lizard tails in the 1680s, but the first major work in this area was Réaumur's memoir on the regeneration of crayfish claws that appeared in 1712 (see Moeschlin-Krieg 1953; Roger 1963:390–96). Réaumur drew the parallel between regeneration and normal generation and was critical of the notion that hidden "germs" might be responsible for the development of the new appendages. Yet, Réaumur noted, all other hypotheses seemed inconceivable as well (Roger 1963:392–94).

The most astonishing development in this area came with Abraham Trembley's discovery in 1741 of a new animal, the freshwater polyp. Trembley, who observed the common green hydra, initially assumed the organism was a plant. Yet the creature's ability to contract when stimulated and to "walk" by attaching successively its posterior and anterior ends to the surface of an object raised doubts in his mind. Trembley then proceeded to cut a polyp in two, assuming that the two halves might live if it was a plant. "However," he admitted, "... I expected to see the cut polyps die" (1744:26). To his surprise, not only did both halves live, but each grew into a new complete animal. As Réaumur, to whom Trembley communicated his findings, expressed this sentiment, "when I saw for the first time two polyps form little by little from the one that I had cut in two, I could hardly believe my eyes; and it is a fact that I am not accustomed to seeing after having seen it again and again hundreds and hundreds of times" (1734–42, 6:liv–lv). Trembley also observed the polyp's normal reproductive method, that is, by budding, a further indication of the plantlike properties of this unusual animal (see Figure 2; see also Baker 1952).

Trembley's discovery of the polyp's ability to multiply *par boutures*, by artificial division, created a tremendous stir among eighteenth-century intellectuals. Réaumur, who announced Trembley's findings and demonstrated his experiments before the Paris Academy of Sciences in 1741, remarked, "I have seen no one who has believed this on the first account he has heard of it" (1734–42, 6:li). The official report of these sessions captures some of the excitement and sense of wonder that must have been present:

> The story of the Phoenix that is reborn from its ashes, wholly fabulous as it is, offers nothing more marvelous than the discovery of which we are about to speak. The chimerical ideas of the palingenesis or regeneration of plants and animals, which some alchemists have thought possible by the assembly and reunion of their essential parts, only tended to restore a plant or an animal after its destruction; the serpent cut in two and said to join together again, only gave one and the same serpent; but here is nature going farther than our fancies. [*Histoire de l'Académie Royale des Sciences*, 1744:33–34]

What was it about the polyp's unusual regenerative capabilities that was so astonishing and so disturbing? The Paris report concludes by recommending that the reader "draw his own consequences" from the polyp's behavior "on the generation of animals, on their extreme resemblance with plants, and per-

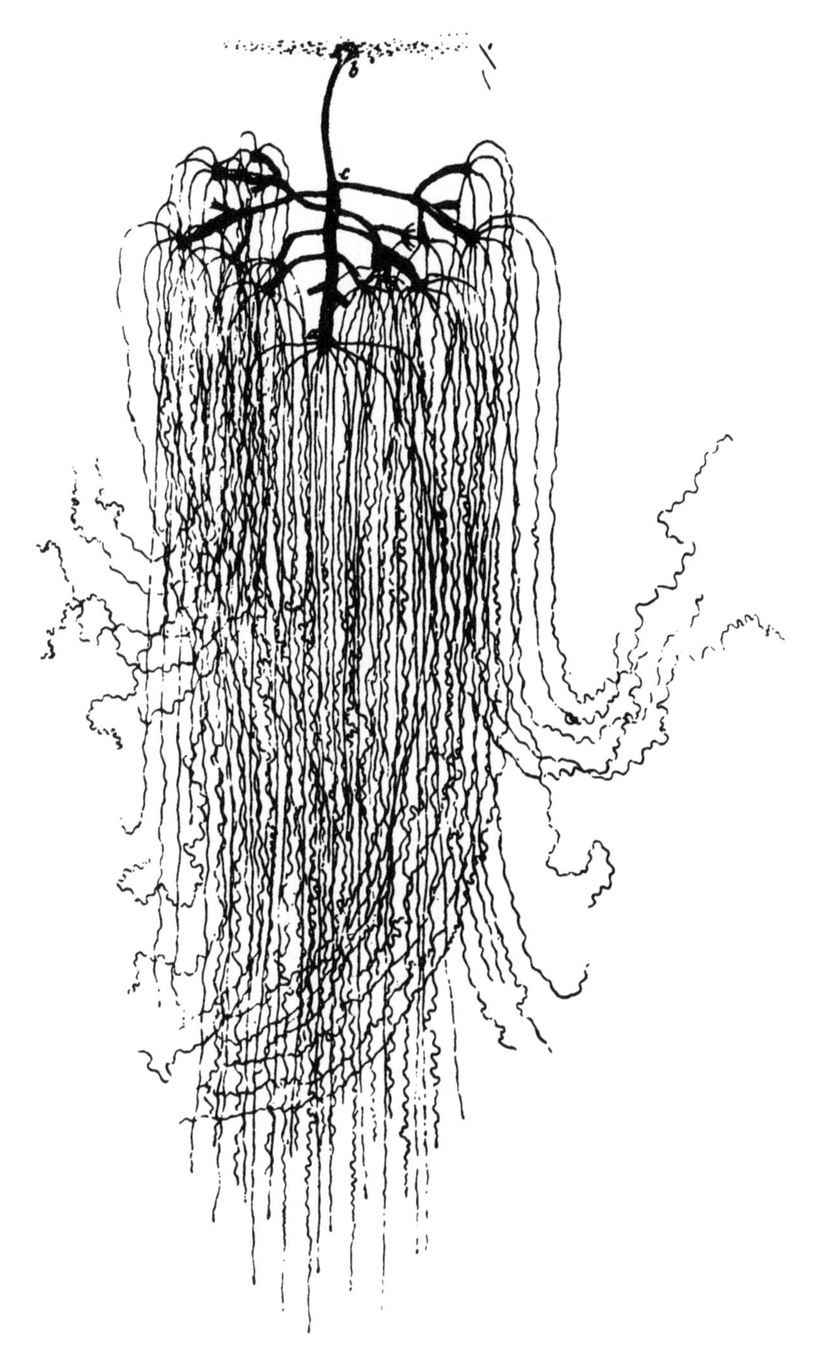

Figure 2. Trembley's illustration of polyps reproducing by budding. (From *Mémoires pour servir à l'histoire d'un genre de polypes d'eau douce*, 1744; courtesy of Cornell University Library)

haps on even higher matters" (1744:35). The polyp's ability to multiply by artificial division raised important questions concerning the divisibility of the soul (Were two new polyp-souls created with two new heads?), the continuity or discontinuity between the plant and animal kingdoms (Was the polyp the intermediate link?), and the nature of animal generation (What implications did regeneration in the polyp hold for normal generation?). At a time when most naturalists believed in preordained generation through the encasement of embryos, the polyp's capacity to create whole new organisms from small pieces of a former polyp was a most unexpected phenomenon. As Aram Vartanian has remarked, "In the pieces of a cut-up polyp regenerating into complete new polyps, Trembley's contemporaries had the startling spectacle of Nature caught, as it were, *in flagrante* with the creation of life out of its own substance without prior design" (1953:388).

Some naturalists responded by proposing that all regeneration is a product of preformed germs contained within the bodies of polyps and other organisms that are destined to develop if an accident occurs. Others saw in the polyp unmistakable evidence for epigenesis. Finally, there were those who simply refused to acknowledge the facts. As Trembley noted, in describing the reaction of the Royal Society to his discoveries, "The singular facts that are contained in the history of these small animals are the admiration of a great many people: but several people have been hesitant to admit them. There are those who have even said that they will not believe it when they see them. Apparently these men have some cherished system that they are afraid of upsetting" (M. Trembley 1943:165–66).

The episode of the polyp illustrates well the close dependence of eighteenth-century theories of generation on philosophical, metaphysical, and religious issues. The polyp was instrumental in the conversion of some to epigenesis (Bodemer 1964), and became a central motif in the materialist theories of epigenesis propounded by Diderot and La Mettrie (see Vartanian 1950). Yet after creating such an initial stir, regeneration in the polyp and in other organisms was quickly subsumed by preformationists like Charles Bonnet under ad hoc explanations based on preexistent part germs. As Roger has noted, it would seem "as if this system . . . was more valuable than the facts themselves" (1963:385).

EPIGENESIS AND ATTRACTION

Maupertuis

Prior to the mid-eighteenth century, the theories of epigenesis that preformationists sought to combat were based for the most part on mechanistic models of particle movement and fermentation. Yet in 1745, when the anonymously published *Vénus physique* appeared, a new challenge to preformationism arose – epigenesis through attractive forces. As Maupertuis, the author of this small treatise, explained, "The Astronomers were the first to feel the need of a new principle to explain the movements of the celestial bodies and thought they had discovered it in these very movements. Since then chemistry has felt the same necessity of adopting this concept, and the most famous Chemists admit Attraction and extend its function farther than had been done by the Astronomers. Why should not a cohesive force, if it exists in Nature, have a role in the formation of animal bodies?" (1745, 1966 trans.:55–56).

Maupertuis proposed that the seminal fluid in both the male and female parents contains particles sent from each part of the body. When these two fluids mix, the proper embryo results by a union of the particles from each part of the parents' bodies. "If there are, in each of the seminal seeds," Maupertuis explained, "particles predetermined to form the heart, the head, the entrails, the arms and the legs, if these particular particles had a special attraction for those which are to be their immediate neighbors in the animal body, this would lead to the formation of the fetus" (p. 56). As an analogy for this formation of the embryo by attractive particles, Maupertuis suggested the *arbre de Diane*, a treelike figure that forms on the surface of water from the mixture of silver, nitric acid, and mercury. Other similar examples of "organized" nonliving formations were known to be abundant in chemical phenomena. Why may these not serve as examples of the way embryos form? "Although these seem less highly organized than the bodies of most animals, might they not depend on the same mechanisms and on similar laws?" Maupertuis asked (p. 55).

Maupertuis opposed preformation on the grounds that it is no easier to explain how all organisms were formed at one time in the past than at each new instance of generation. "What has natural science lost by the idea that animals are formed successively? For God, is there any real difference between one moment in time and the next?" (p. 42). There were two classes of evidence that Maupertuis felt made preexistence impossible. The first was Harvey's observations on the gradual formation of embryos. Upon opening the uteruses of does shortly after mating, Harvey found no tiny embryo developing. Rather, he found no change at all for some time and then only a gradual development. Second, Maupertuis felt that neither ovist nor animalculist preformation could be reconciled with the fact that offspring show resemblance to *both* parents. Hybrid organisms are an even more striking example of this. Thus, the embryo must be a composite of generative material from both the mother and the father. Consequently, Maupertuis concluded, we should return to the ancient notion, promoted also by Descartes, that the embryo forms out of seminal fluids from both parents.

Maupertuis did not explain his theory of generation in any further detail, either with regard to the seminal particles or the attractive force. Yet because of the phenomena of resemblance and heredity, Maupertuis felt that his was the only viable alternative to preformation. Maupertuis's theory was not widely accepted, and it met with much initial criticism. Réaumur, in a work devoted to techniques of incubation for raising domestic chickens, attacked in particular Maupertuis's use of the concept of attraction in his explanation of generation. Rejecting first Descartes's theory on the grounds that blind fermentation of seminal material could never be capable of forming a living organism, Réaumur brought similar arguments to bear against Maupertuis (without mentioning the anonymous author by name). It had become fashionable, Réaumur charged, for occult qualities like attraction, which had been banished from philosophy, to be used once again in physics and to be extended into other areas as well, even the formation of the fetus. Yet, Réaumur countered,

> how will attractions be able to give to such and such a mass the form and structure of a heart, to another that of a stomach, to a third one that of an eye, and to another that of an ear? How will they frame other masses into vessels, valves, etc. [A]ll their tendency will amount barely to the reunion of the similar parts into solid masses. What law of attraction shall one imagine for the making of that small bone of the ear, whose figure makes it to be called the stirup? How shall so

> many different organs be placed and assembled in their proper order? We see with the most glaring evidence, that in order to arrive at the formation of so complicated a piece of work, it is not enough to have multiplied and varied the laws of attraction at pleasure, and that one must besides attribute the most compleat stock of knowledge to that attraction. [1749, 1750 trans.:463]

Without guidance, how could the seminal particles become organized into a developing embryo? Such questions were impossible to answer on strictly mechanistic grounds, even when attractive forces were invoked.

Maupertuis's response was to move away from a rigidly mechanical view of matter. Hinted at in a concluding "Doubts and Questions" section of the *Vénus physique* and developed more fully in his *Systême de la nature* (1751), Maupertuis's solution was to attribute to the smallest particles that make up living organisms a capacity of intelligence and memory. By "remembering" their former locations and by possessing an instinct to unite, the seminal particles would be able to arrange themselves into an embryo in the proper fashion. "A uniform and blind attraction," Maupertuis admitted in the *Systême de la nature*, "diffused in all the particles of matter, would not be able to serve to explain how these particles arrange themselves to form the body whose organization is the most simple. . . . it is necessary to have recourse to some principle of intelligence, to something resembling what we call desire, aversion, memory" (1751, 2:146–47).

In proposing this solution to the source-of-organization dilemma, Maupertuis rejected the traditional mechanistic view of passive matter and adopted instead a Leibnizian notion of matter as fundamentally active (see Roger 1963:479). Adding a psychical quality to dead particles of matter, Maupertuis was able to answer the objections of those who, like Réaumur, saw epigenesis by blind attraction as an untenable explanation. Maupertuis's solution was to offer an alternative not only to preformation but also to the rigidly mechanistic philosophy on which it was based.

Buffon

During these same years, another antipreformationist theory of generation appeared in the second volume of Buffon's *Histoire naturelle* (1749). Buffon's theory was based on a distinction between two kinds of matter, organic and brute, and on an identification of the reproductive process with nutrition. During an animal's life, organic particles are taken up in food, are separated from brute matter during digestion, and are used by each part of the body for growth and maintenance. When adult size is reached and more organic particles are taken in than are needed, the surplus particles are collected in reservoirs and become the seminal liquid. These superfluous nutritive particles come from all parts of the organism's body as "representatives" of those parts. During reproduction, these particles come together to form the new organism, which, in the case of sexual reproduction, results from a mixing of seminal material from both parents.

Buffon performed a number of experiments with his colleague John Turberville Needham to show that a semen exists in females as well as in males. They even thought that they had observed animalcules in female semen akin to the male spermatozoa. Buffon did not regard these animalcules as true organisms, however, but rather as "mere machines" that result from the chance combinations of organic particles in the seminal fluids.

Buffon rejected the preexistence of germs as an explanation for generation. Such an account of reproduction "is not only admitting that we do not know how it is accomplished but also renouncing all desire to conceive of it" (1749–89, 2:28). Instead, Buffon proposed that the formation of the embryo takes place after the seminal fluids from both parents mix. This process is accomplished by two agents, the organism's *moule intérieur* ("internal mold") and the existence of special "penetrating forces." The embryo forms immediately upon the mixing of the fluids, then proceeding to grow by the assimilation of organic particles through nutrition.[9]

Buffon's concept of the internal mold was intended to be his answer to both preexistence and mechanical epigenesis. "In the same manner as we can make molds by which we give to the exterior of bodies whatever figure we please," he explained, "suppose that Nature can make molds by which it gives [to organic bodies] not only exterior figure but also internal form; would this not be a means by which reproduction could be brought about?" (1749–89, 2:34). Could we not suppose that there is some kind of organizational matrix that is responsible for the proper assembly of particles during both growth and reproduction? When an animal or plant reproduces, Buffon

Figure 3. Searching for animalcules in the female semen. *Left to right*: unknown figure, Daubenton, Needham, Buffon. (From Buffon, *Histoire naturelle*, vol. 2, 1749; figures identified by Punnett 1928)

maintained, the internal mold is passed on to the offspring, there becoming the basis for growth. When the organism matures, excess organic particles are collected and through them the internal mold is passed on again to new offspring, completing the cycle of growth and reproduction.

In order to explain how the internal mold accomplishes these tasks, Buffon invoked certain "penetrating forces," analogous to the forces that govern gravity, magnetic attraction, and chemical affinity. "It is apparent . . . ," Buffon claimed, "that forces exist in Nature, such as gravity, which relate to the exterior qualities of bodies, but which act on the most intimate parts and penetrate to all points. . . . for, in the same manner as the force of gravity penetrates the interior of all matter, the force that pushes or attracts the organic parts of nourishment penetrates into the interior of the organized body" (pp. 45–46). The penetrating forces cause the organic particles absorbed from food to be properly assimilated to the internal mold in each part of the organism's body. Just so, with regard to reproduction, "May not a force similar to that which is necessary to . . . produce growth, be sufficient to bring about the reunion of these organic particles, and indeed to assemble them into an organized form similar to that of the body from which they are derived?" (p. 62). The penetrating force ensures that the representative organic particles, and the internal mold they carry, become properly assembled in the new embryo.

Buffon's theory of generation shared several of the difficulties encountered by Maupertuis. Both saw the seminal fluids as being made up of particles acting as representatives from the parts of the parents' bodies. Yet neither fully explained in what this similarity actually consists. Buffon's internal mold was intended to account for this, yet he never disclosed exactly what the internal mold is. Both Maupertuis and Buffon invoked attractive forces to guide the formation of the embryo out of the seminal fluid, yet both left themselves open to Réaumur's charge that blind attraction could never accomplish this task (although, as we saw, Maupertuis attempted to solve this difficulty through "intelligent" particles). It is significant, with regard to their utilization of attractive forces, that both Maupertuis and Buffon were early supporters of Newton on the Continent.[10] The freedom that Newtonian mechanism offered to biology – the addition of force to matter and motion as the fundamental categories of explanation – played an important role in embryology as well as in other areas of physiology.

Needham

A third theory of generation contemporaneous with those of Maupertuis and Buffon was put forward by John Turberville Needham. Following his laboratory collaboration with Buffon, Needham proposed his own explanation of generation in a paper published in the *Philosophical Transactions of the Royal Society* for the year 1748 and in an expanded French version, which appeared in 1750 (*Nouvelles Observations microscopiques*). Needham postulated that there exists a universal semen from which all generation occurs. At the lowest level are formed the spermatic animalcules and the animalcules found in infusoria and in pond water. (Contrary to Buffon, Needham classified all of these as living organisms.) The constant combination, separation, and recombination of organic matter are all produced, Needham claimed, by a universal vegetative force, the source of all life activity. This vegetative force may further be broken down into two forces, one of expansion and one of resistance, that are constantly balanced against one another (1750:221 n.). Through a continual tendency to expand, checked by a cohesive force of resistance, living matter was thought by Needham to produce the observed phenomena of vitality.

Needham's system had its greatest impact on his contemporaries in its support for spontaneous generation. Needham claimed that he had found animalcules spontaneously forming in boiled meat infusoria. Decried as impious and materialist by some, Needham's views led others, notably Lazzaro Spallanzani, to attempt experimental refutation. Spallanzani, himself a preformationist, believed his work proved that Needham was in error, and that the appearance of animalcules in infusoria was the result of Needham's having used insufficient sterilization techniques (Spallanzani 1765).[11] Needham responded in 1769 by publishing an annotated French version of Spallanzani's book in which Needham claimed that the excessive heat used by Spallanzani had destroyed the vegetative force and had impaired the elasticity of the air within the flasks, thus rendering them incapable of producing animalcules. "What confusion, what obscurity prevails in his notes to my microscopical observations! What monstrosities in his thoughts!" Spallanzani complained to Bonnet (cited in Dolman 1975:556). Resolved to silence Needham once and for all, Spallanzani devoted himself to further experiments, producing his masterful *Opuscoli di fisica animale e vegetabile* (1776). His work was warmly received by Bonnet and other preformationists, who uniformly opposed spontaneous generation.

The three systems of Maupertuis, Buffon, and Needham represented a combined challenge to preexistence theories at midcentury. The similarities among their views may be due in part to the fact that they were in close communication during the late 1740s. Needham and Buffon collaborated in some of their experiments; and, according to Needham (1748:633), Buffon and Maupertuis often discussed the subject of generation following the publication of the *Vénus physique* (1745). Yet the similarity in these theories also represents a common dissatisfaction with the notion of preexistence, with its inability to deal with the phenomena of resemblance and hybrids, and, more especially, with the seeming absurdity of the idea of *emboîtement*. Furthermore, all three turned to the concept of attractive forces to resuscitate epigenesis following its overly mechanistic formulation in the fermentation models of Descartes and others.

This challenge was clearly felt by preformationists and was met by a revival of preexistence theories in the work of Haller, Bonnet, and Spallanzani. These three carried the theory of preformation to its most fully articulated level, which was also its last stand before the pendulum swung once again to epigenesis in the closing decades of the eighteenth century. Haller's support for the theory was strengthened by the fact that he had believed in epigenesis before his conversion to preformation in the 1750s. To the story of Haller's evolving beliefs, and their relationship to the contemporaneous events already discussed, the following chapter is devoted.

2

Haller's changing views on embryology

During his lifetime Haller held three distinct theories concerning embryological development. As a student of Boerhaave, he adopted his mentor's belief in spermaticist preformation. Then, in the mid-1740s, he switched his support to epigenesis. Finally, in the 1750s, Haller converted back to preformation, presenting a detailed account of preexistence in the maternal egg in his first major embryological work, *Sur la formation du coeur dans le poulet* (1758a). Haller's changing views can be most clearly understood when seen within the context of both the philosophical issues and the biological problems that occupied eighteenth-century embryologists.

Haller's early education took place in his native Switzerland. After spending a year studying medicine in Germany at the University of Tübingen, Haller moved in 1725 to Leiden to study under the famed Boerhaave, obtaining his medical degree in 1727.[1] Boerhaave's influence on Haller was formative in more than just physiology. As we shall see more fully in Chapter 4, Boerhaave's philosophical views on science were reflected in significant ways in Haller's work.

One of the doctrines Haller adopted from his teacher was Boerhaave's belief in animalculist preformation. Boerhaave taught that the rudiments of the embryo preexist in the animalcules found in the male semen, developing into a complete organism in the female: "The father gives the embryo and first rudiments of life, the structure of the body being already created and prescribed in the animalcules engendered in all animals. . . . The mother receives the living rudiments from the father, and retains and nourishes the same, maintaining therefore a dwelling place for the fetus" (Haller 1739–44, 5, pt. 2[1744]:485, 487; see note 3). Haller's belief in this doctrine appears to be that of a student taking on the doctrines of his teacher, since he never published any major statement defending Boerhaave's views. Yet in a poem, written in 1729, Haller remarked, in a section devoted to illustrating the wisdom of the Creator, "Indeed, in the semen already, before life breathes,/ The ducts are already formed, that first the animal uses"[2] (Hirzel 1882:58, lines 345–46). Haller's preformationist leanings were also expressed in a publication on monstrous births, which appeared in 1739. Here Haller opposed the views of Lémery, who believed that monsters are formed by the accidental union of previously separate organisms (or parts of organisms). Haller countered with the claim that monstrous structures are preformed, just like all others, in the first rudiments of the fetus (1739:27–34; see also Sturm 1974:100–101).

The most straightforward evidence for Haller's belief in spermaticist preformation can be found in remarks he made after he converted to epigenesis. In an edition of Boerhaave's lectures that Haller prepared and edited (*Praelectiones academicae in proprias institutiones rei medicae*, 1739–44), Haller offered critical comments in footnotes.[3] In one of these, which surveys various theories of generation, Haller reports that "Others, finally, among them [my] preceptor [Boerhaave], say that man preexists solely in the *vermiculus*. . . . But though I have been brought up on this theory, I am nevertheless forced to abandon it" (Haller 1739–44, 5, pt. 2[1744]:499, n.*g*; trans. Adelmann 1966, 2:894, 895).

After Haller left Leiden, he traveled to England and through Europe, and lectured on anatomy in his native Switzerland. In 1736, he received an appointment as Professor of Anatomy, Surgery, and Medicine at the newly founded University of Göttingen. He was to remain at Göttingen for almost twenty years, there establishing one of the foremost medical schools on the Continent (second only to Leiden). It was at Göttingen that Haller first became clearly interested in theories of generation and converted to the side of the epigenesists. This took place in the 1740s and was prompted by Trembley's discovery of the remarkable properties of the freshwater polyp.

HALLER'S THEORY OF EPIGENESIS

Trembley's report of reproduction by artificial division in the polyp led several naturalists to search for other organisms possessing this same property. One result of this was Charles Bonnet's discovery in 1741 that certain freshwater worms are also able to regenerate two organisms when one is cut in two. Bonnet published his observations on these worms in his *Traité d'insectologie* (1745), where he also announced his discovery of parthenogenesis in the aphid.

Haller's response to these discoveries was to convert to epigenesis and to use the example of regeneration in the polyp as a model for epigenetic development in all organisms. In a review of Bonnet's *Traité d'insectologie* anonymously published in 1746,[4] Haller noted that before Trembley's and Bonnet's discoveries, almost everyone believed in either ovist or spermaticist preformation. Yet, "This system, which has been so well recognized, is coming now to its end. One is forced to avow, after the observations made on the polyp, that several animals form themselves from heads, from arms, from organs of all types . . . where one could never suspect a miniature to have existed" (1746a:188). Thus, Haller maintains, "Eyes more attentive and minds free from systems are beginning to be persuaded that the most perfect animals are born in approximately the same manner [as the polyp], that their formation is successive and that there was never a plan in which their parts were designed in miniature" (p. 188). Such phenomena as the formation of the heart in the chicken, which appears at first to be a simple tube with no resemblance to the resulting four-chambered heart, and as the apparent production of the fibers and membranes of the embryo from the coagulation of viscous fluids, point to a mechanism of formation similar to that which occurs in the polyp. "I am persuaded," Haller concludes, "that after several more years of observation, one will find that animals and thus plants are engendered from a fluid that thickens and organizes itself little by little, following laws that are unknown to us but which the eternal Wisdom has rendered invariable" (pp. 188–89). Gradual development from unorganized fluid will become, according to Haller, the paradigm for the formation of all organisms, based on the model of the polyp.

Haller outlined similar reasons for his switch to epigenesis in his edition of Boerhaave's lectures, cited earlier, where he first announced his conversion to epigenesis. In the same footnote in which he discussed Boerhaave's support for spermaticist preformation and his own rejection of this theory, Haller points to the originally fluid state of all embryos, the changes the heart undergoes during development in chick embryos, and the fact that some parts exist in the embryo that do not end up in the adult (for example, structures in the caterpillar that are not in the butterfly) as supporting arguments for epigenesis. "Finally," Haller declares, "experiments performed on plants and animals make escape almost impossible. . . . It now seems to me most difficult to tell (what the advocates of evolution[5] [preformation] must nevertheless tell) . . . how the same parts which in the morning were little cuttings of a spine, a stomach, or a head, by afternoon change into true heads and whole stomachs. . . . There is no hope of an explanation here, unless you concede that the polyp's fibers . . . [can] shape themselves into an animal, of which there was no likeness in these fibers" (1739–44, 5, pt. 2[1744]:504–6, n.*g*; trans. Adelmann 1966, 2:898–99).

Although Haller presents persuasive arguments in favor of gradual development, in neither of these two discussions does he explain how epigenesis really works. In his edition of Boerhaave's lectures, Haller concludes simply that "it therefore seems probable to me that the parts of animals are successively generated from a fluid in accordance with definite laws" (p. 506, n.*g*; trans. Adelmann 1966, 2:899). And although he rules out fermentation as being too violent to produce anything organic, he does not define these laws any further. Similarly, in his review of Bonnet's *Traité d'insectologie*, Haller remarks that development follows "laws that are unknown to us but which the eternal Wisdom has rendered invariable" (1746a:189). Yet what these laws are and how they operate are not discussed.

In 1747, in the first edition of his *Primae lineae physiologiae*, Haller offered a mechanism for how epigenetic development proceeds.[6] Rejecting both development guided by the soul and the preexistence of germs, Haller asks "Or indeed [is it] an attractive force that first gathers the viscous liquid into filaments, [then] these into fibers, membranes, vessels, muscles, bones, forming [finally] the limbs? This indeed seems to be the more probable" (1747:478). Furthermore, if one asks how this process of construction "which is so wise, so constant, so varied for particular purposes" is carried out, the answer must be, Haller contends, that it is "Undoubtedly by divine laws, which in their proper manner order spicules of ice, crystals of salts, particles and sheets of metals . . . to be joined together" (pp. 478–79). Similarly, one can point to the formation of moss spores and of the threads of flax or cotton, progressing on up to the liquids of plants, the gluten of the simplest animals, and finally the fibers and membranes of the most noble organisms. All are formed analogously through forces of attraction; all "can be produced only through these laws, from this material, under these conditions" (p. 479).

Haller does not outline any further his notion of epigenetic development through attractive forces. His model is one of an organized solidification of parts from the original embryonic fluid that produces a hierarchical organization of more and more complex structures. All of this is under the guidance of an attractive force that can be seen to operate analogously in inorganic and organic matter. Yet nowhere does Haller explain how all of this takes place. Solidification of fluids could certainly produce the structures of the embryo, but why does it? What causes the process of development to proceed as it does? What directs the operation of the attractive force in the gradual formation of tissues and organs? These are questions that Haller does not discuss. Yet they must be answered, as Réaumur argued in his critique of Maupertuis's model of development by attraction, for any epigenetic theory fully to explain development. As we shall see, Haller's realization in subsequent years that questions such as these must be dealt with played a crucial role in his conversion back to preformation.

In his three major discussions of epigenesis, Haller outlined four principal categories of evidence that necessitated, in his opinion, belief in gradual development. The most important of these, and the one that prompted his own conversion to epigenesis, was reproduction by artificial division in the polyp (and other similar phenomena of regeneration). Haller also cited the apparently original fluid state of the embryo as precluding the preexistence of organized structures, and pointed to the radical changes that some organs, especially the heart, undergo during development. If the four-chambered heart were preformed, how could the chick embryo's heart look like a simple tube in early stages of incubation? Finally, in the *Primae lineae physiologiae*, Haller argued that the fact that organisms, especially hybrids, resemble both parents rules out the possibility that the embryo is preformed in either the mother or the father. It will be interesting to note what role these arguments played, if any, in Haller's subsequent conversion to preformation and how his new theory was able to account for them.

BACK TO PREFORMATION

The most common secondary-source account of Haller's views on embryology is that he converted from epigenesis to preformation because of his own experiments on chicken eggs.[7] Haller himself gives credence to this view in his *Sur la formation du coeur dans le poulet*, where he states, "I have shown sufficiently in my works that I leaned toward epigenesis and that I regarded it as the opinion that conformed the most with experience. But these matters are so difficult, and my experiments on the egg are so numerous, that I propose with less repugnance the contrary opinion, which is beginning to appear to me to be the most probable. The chicken has furnished me with reasons in favor of development [unfolding from preexisting parts]" (1758a, 2:172). Similarly, in a collection of papers and essays that contains his review of Bonnet's *Traité d'insectologie*, in which he had supported epigenesis, Haller adds in a footnote, "at that time I was inclined toward the gradual formation of animals. But more mature observations, particularly those by observant eyes on the formation of chickens, have brought me back since then to evolution [preformation]" (1772b:298 n.).[8]

Haller did indeed perform an extensive series of investigations on incubated chicken eggs during the summers of 1755, 1756, and 1757; and it was in 1757 that Haller first clearly stated his belief in preformation. Yet I must agree with Roger that "it would be difficult to affirm that the conversion of Haller to preexisting germs had no other real cause than his observations" (1963:709).[9] A closer look at the years 1747 to 1757, that is, between Haller's last statement in support of epigenesis and the first declaration of his belief in preformation, reveals that Haller was deeply concerned with questions relating to generation. He was not simply a confirmed epigenesist who switched suddenly to preformation upon his observations on chicken eggs.

Reaction to Buffon's theory of generation

Perhaps the most significant influence on Haller's embryological views during this period was the appearance of Buffon's theory of generation in 1749. Haller's response to Buffon's novel ideas can be seen most clearly in a preface Haller wrote to the second volume of the German translation of Buffon's *His-*

Figure 4. Albrecht von Haller in 1757. (From the frontispiece to *Elementa physiologiae corporis humani*, 1757–66)

toire naturelle. This volume was published in 1752, although Haller's preface appeared as a separate publication in French, titled *Réflexions sur le systême de la génération de M. de Buffon*, a year earlier. Haller also reviewed Buffon's *Histoire naturelle* (volume 2) in the *Bibliothèque raisonnée* in 1751.[10]

In his preface and his review, Haller presents detailed summaries of Buffon's views. Yet criticisms also abound; halfway through his preface, Haller remarks, "Here I cannot agree with the shrewd and learned M. Buffon, and the eternal power of truth calls me away from his opinion. A multitude of objections that present themselves all at once to my mind quarrel for the privilege of presenting themselves first" (1752a:viii). Haller's critical comments, as presented in both the preface and the review, can be grouped into three categories: the existence of the female semen, the question of resemblance to parents, and the operation of the internal mold.

Haller simply rejects the notion of a female semen on the basis of his own anatomical knowledge. There is no proof for its existence; and, furthermore, Buffon's own arguments in support are fallacious, Haller maintains. With regard to the issue of resemblance, Haller again challenges Buffon's claims. One of Buffon's major arguments in support of his two-semen theory was that children resemble both parents. But Haller, again on the basis of his own anatomical studies, denies that children ever really resemble their parents. In fact, he claims, there are no two individuals who are ever completely alike anatomically; the structure and placement of the nerves, blood vessels, and other organs vary immensely from one person to the next. Furthermore, Haller argues, if children are formed from particles sent from all parts of the parents' bodies, how can deformed parents produce normal children?[11]

Haller's third major objection to Buffon's theory was the most important with regard to Haller's own changing views on generation, for it raised the whole issue of how forces are involved in embryological development. Haller was unable to see how Buffon's *moule intérieur* could possibly organize all the organic particles into a complete organism. Even if we assume that all the particles sent from all the parts of the body are in the seminal fluid, how are they put together? "M. Buffon has shown no cause," Haller claims, "that brings them in order, that unites the eye particles of the father with the eye particles of the mother. . . . There is missing a building master, who lays down the thousand single molds of the different parts of the large aorta in a correct row according to the length of the body, and who, with a word, would construct the scattered microscopical parts of the body according to the wonderful plan of the human body, who would make sure, that never could an eye stick to a knee, or an ear to a forehead" (1752a: x). How could all of these particles come together to form an organized structure without such a "building master"?

Buffon, we can recall, invoked the concept of a "penetrating force" to accomplish this task of organization. Haller counters, however, "I do not find in all of nature the force that would be sufficiently wise to join together the single parts of the millions and millions of vessels, nerves, fibers, and bones of a body according to an eternal plan. . . . M. Buffon has here the necessity of a force that seeks, that chooses, that has a purpose, that against all of the laws of blind combination always and infallibly casts the same throw" (p. xi). But could such a force possibly exist? Haller's answer is no.

The issue of forces and how they operate in development figures centrally in the concluding section of Haller's preface, where he discusses the question of whether the ability of matter to produce living creatures by itself, via forces, would eliminate the need for a Creator. Such charges had been made against Buffon's theory by theologians at the Sorbonne. Haller's response is simply that matter and its forces by themselves are not responsible for development, that on their own they would be incapable of producing a new organism. The force of attraction, Haller admits, figures in the formation of salts and crystals, in making snowflakes and the *arbre de Diane*,[12] in freezing ice, and in producing metals. "Should it be so far," Haller asks, "when one crosses over from the above-named seedless constructions [salts, snowflakes, etc.] to the simplest animals, which in all their parts are a similar and uniform jelly? . . . Where does the power of universal laws stop? Where is the boundary, within which they form, and outside of which they can build no more?" (1752a:vi). Haller's answer is that it is God who directs and limits the operation of universal laws and forces. Matter alone is powerless, yet when given forces by God it is capable of producing all of the natural phenomena we observe. "If matter has forces that allow it to build things, it does not possess them blindly. They are surrounded by eternal limits, and build always perfectly not mechanical equals but something similar, something that is prescribed in an inviolable plan: but with a diversity that excludes the constraint of blindly working matter" (p. xv).

There are two key arguments involved in this last passage that Haller used to prove that forces alone could not govern the world. First, Haller believed that matter possesses no abilities whatsoever on its own, without these having been designated by God. As he explains in his review of Buffon's theory, "Matter does not possess these abilities itself. It could be without gravity, without elasticity, without irritability; added qualities, but essential to the structure of plants and animals. These qualities do not take part in its essence; they are foreign to it; they are not common to all parts of matter. . . . A first cause has thus allotted to different classes of matter abilities and forces calculated according to a general plan, and it is there that we recognize the hand of the Creator" (1751b:86). Matter can do absolutely nothing without the forces that God gives to it. Thus, that forces of matter produce natural phenomena does not threaten the existence of a Creator.

Haller's second argument rests on the fact that both variety and constancy of form exist among living organisms. If matter formed living creatures blindly, why would species be preserved? "If a mere attractive and repulsive force in the semen built a man or a deer; if this occurred by chance, why does there never develop from this . . . matter . . . instead of a man, a monkey, which has so many similarities with man?" (1752a:xv).[13] Haller's answer again is that it is God, not matter, that is ultimately responsible for the formation of organisms. A great deal of variety exists in specific structures within living creatures. Reproduction does not produce an exact duplicate but rather an organism that is similar in its essential details. "This invariable constancy of species, which permits a slight variety but which never deviates in these essential characters, appears to me to be one of the most evident proofs of the hand of the Creator" (1751b:87). Consider, Haller suggests, the case of a carnation. The matter out of which it is formed may produce a carnation of a different color or even of a different structure, but it never produces a tulip. "It is God who permits a double carnation to be made, different from the crimson carnation native to the lakes of the Alps, but always a carnation, which in a thousand generations will never produce anything but a seed of a carnation. All is not chance, otherwise the carnation would become a tulip: all is not necessity, otherwise the carnation would remain always such as was the first carnation" (1751b:87). God is the mastermind of the developmental process, not matter alone.

Haller's conclusion on this issue of the relative roles of forces and God in development is simply that "The invariable production of always similar, always divinely constructed animals, appears to be too great for the simple forces that produce . . . a salt crystal" (1752a:xvi). The boundary between the production of simple structures like crystals and the formation of living organisms is drawn by God. Yet is this not the exact opposite of what Haller had argued only a few years earlier in support of his theory of epigenesis based on attractive forces? There he had likened the laws governing development to those through which crystals and metals are formed. Now he maintains that development is too much for these simple forces of attraction to be able to accomplish by themselves.

The central issue in all of these discussions of forces is guidance. Haller never questioned the idea that simple forces of matter are active in development. His concern was with *how* they act to produce their effects, since he did not believe that they could accomplish the task of development unaided. God, Haller believed, directs and limits the operation of forces in producing the developing embryo. Yet how this guidance is manifested, that is, what the mechanism is through which God controls development, is not made clear by Haller in his discussions of Buffon's theory.

Haller's ideas on generation had clearly altered since his earlier statements in support of epigenesis. No longer could he postulate that the attractive forces of matter by themselves produce fibers, membranes, and organs from an originally unorganized liquid. He was not yet a preformationist, yet he was clearly at this time in a transitional stage, moving away from his earlier epigenetic views, beginning to question the relationship between forces of matter and development. On the one hand, he still believed that development occurs *by means of* forces of matter; on the other hand, he was reluctant to allow that these forces *alone* could be responsible for generation. God's wisdom must somehow direct development. But how forces of matter are guided by God and exactly how they operate on the material of the embryo was still not clear to Haller.

Haller dispensed with the views of Needham even more cursorily in his discussions of Buffon's work on generation. Needham's vegetative force was just the kind of force Haller could not allow. "And yet," he declared, "I am not troubled by the very remote proof of these building forces. Truth is like a correct calculation, established on all sides; everything must help support its structure; only error falls apart as soon as one takes away its only support, because all else conflicts against it" (1752a:xiv). Similarly, he remarked at the conclusion of his preface, "We can thus wait tranquilly to see if the experiments of the learned will confirm or contradict the vegetative and animating forces of M. Needham. They will lead us always nearer to the truth, and this to God" (p. xvi). By the time Haller's preface was reprinted in his *Sammlung kleiner Hallerischer Schriften* (2nd edition, 1772b), Haller was able to add footnotes referring to Spallanzani's experimental refutation in 1765 of Needham's views on spontaneous generation (see Chapter 1). And in 1767, when Haller included a Latin version of this preface in his *Opera minora*, all references to Needham's views were simply deleted.[14]

Haller rejected both Needham's vegetative force and his promotion of the idea that living organisms could spontaneously generate from unorganized material. As we shall see in later chapters, the issue of spontaneous generation was intimately tied for Haller to the larger question of the nature of generation in all organisms, and it played a role in his rejection of Wolff's theory of epigenesis. It may be that the parallel, drawn explicitly by Needham, between epigenetic development in higher organisms and the spontaneous generation of lower forms of life contributed to Haller's growing unease with the theory of epigenesis. Spontaneous generation, if true, would certainly call into question the role of God in the creation of new organisms. And the formation of higher organisms, as described in both Needham's and Buffon's theories, bore a dangerous resemblance to the formation of animalcules in Needham's infusoria.

Force and the concept of irritability

During these same years (late 1740s to early 1750s), Haller was developing his ideas on another force concept, that of irritability. Haller's first reference to the notion of irritability may be found in a footnote in his edition of Boerhaave's lectures, *Praelectiones academicae*. Here, commenting on Boerhaave's description of the systolic motion of the heart, Haller remarked that since movement persists in hearts in animals that have recently died, the heart must beat from some "unknown cause" that "lies hidden in the fabric of the heart itself" (1739–44, 2[1739]:129, n.*i*). Although Haller made similar remarks in the fourth volume of the *Praelectiones* (4[1743]:586, n. *a*) and in his *Primae lineae physiologiae* (1747:51; 1751c:252), his first major exposition of the theory of irritability and sensibility was his famous paper, "De partibus corporis humani sensilibus et irritabilibus," which was presented in 1752 to the Royal Society of Sciences of Göttingen and published in its journal the following year. Here he argued that irritability, the capacity of muscles to contract, is completely separate from sensibility, that is, sensation. Haller ascribed sensibility to the nerves and irritability to the muscles, which possess an inherent force that causes them to contract when stimulated. Irritability, called in later publications the *vis insita*,[15] "originates in the very fabric of the irritable part" (1752b:134; see also Haller 1936). Both the existence of involuntary motions in live animals that are entirely independent of the will, and the contractions that occur in muscle tissue in a dead animal or in muscles separated from the body, point to the independent property of irritability in muscles. "What therefore should hinder us," Haller proclaimed, "from granting irritability to be that property of the animal gluten in the muscular fiber, such that upon being touched and provoked it contracts, to which moreover it is unnecessary to assign any cause, just as no probable cause of attraction or gravity is assigned to matter [in general]. It is a physical cause, hidden in the intimate fabric, and discovered through experiments, which are evidence enough for demonstrating its existence, [but] which are too coarse to investigate further its cause in the fabric" (1752b:154). Irritability is a force inherent in a particular type of matter (animal muscle tissue) that operates automatically under proper conditions of stimulation. Like gravity, irritability can be deduced from phenomena, yet its ultimate cause lies, for the time being at least, beyond the reach of experiment and observation.

During the 1750s, Haller further developed his ideas on the nature of irritability, particularly in relation to the theories of Robert Whytt, who believed that muscular contraction is based on the soul, and in relation to the materialist position of La Mettrie. From these various controversies, there emerges a picture of Haller attempting to steer a middle course between animism on the one hand and materialism on the other.[16]

Figure 5. Researching irritability and sensibility. (From the frontispiece to Haller, *Mémoires sur la nature sensible et irritable des parties du corps animal*, 1756–60; courtesy of the Francis A. Countway Library of Medicine, Harvard Medical School)

Haller believed that irritability is a force of matter not based on the soul or any other immaterial force, yet he was also concerned to show that matter could not on its own possess the power of irritability. This ability is given to matter by God: "I piously acknowledge that God is the mover of all nature. For this reason, neither the elasticity of expanding air, nor the weight of a stone, nor the effervescence of acids mixed with alkalis, nor the contraction of a dissected muscle ought to be attributed to incorporeal forces. God gave to bodies an attractive force and other forces, which once received are exercised" (1757–66, 7:xii). God is the source of irritability, gravity, and the other forces that matter possesses.

Haller's ideas on the nature of irritability are important in two ways for our understanding of the development of his views on embryology. First, they offer us an example of Haller's philosophical beliefs about biology and about how forces operate in biological phenomena. Forces are not inherent in matter but are given to matter by God. They operate on a mechanical basis, however, once bestowed upon matter, automatically acting under specified conditions.

This view of forces is not inconsistent with the kind of epigenesis Haller had suggested in 1747, when he claimed that development proceeds via attractive forces and "divine laws." Yet, as both Roger (1963:709–10) and Gasking (1967:114) have pointed out, the kind of epigenesis Haller could accept, based only on mechanical forces, was simply not viable. How could such forces account for the organization of the embryo? What was to serve as Haller's "building master"? In his statements supporting epigenesis and in his discussions of Buffon's theory, Haller turned to God as the source of this guidance. Yet nowhere does he outline *how* God directs the developmental process, how the automatically acting mechanical forces in matter are marshaled to produce the proper offspring.

Haller's work on irritability and his reflections on the implications of the Buffonian system of generation forced him to come face to face with the problem of reconciling his belief in the world as a Divine Creation with mechanical explanation. In criticizing Buffon's theory, Haller began to realize that simple forces of attraction were not sufficient to account for the developmental process. But what forces would be sufficient? How can one explain development on a mechanical basis and still retain a place for the guidance of God?

In relation to this question, Haller's work on irritability provided him with more than just a model for how biological forces operate. This is the second, and perhaps most important, aspect of the significance of Haller's irritability theory for his changing views on embryological development. For after Haller converted to preformation, irritability became the central force responsible for initiating the process that turns preformed and rudimentary parts into a fully developed organism. Irritability provided the final key to the mechanism of God's guidance. The embryo, created by God at the beginning of the world and encased within its parent, had only to be stimulated by the male semen for the inherent irritability of its heart to respond and begin to beat, thus initiating the unfolding process. But this is getting ahead of my story, for it was during an important series of experimental investigations that Haller's conversion to preformation took place.

Observations on chicken eggs and conversion to preformation

In writing his preface on the Buffonian theory of generation, Haller realized that his criticisms of Buffon's ideas carried over into his own previous work on the subject. As he relates, "My objections go against myself . . . and send me back to the difficult necessity of searching for myself" (1752a:xiii). Within a year, Haller had begun a series of experiments on sheep with one of his students, Johann Christoph Kuhlemann, who published the results of these investigations as his doctoral dissertation (1753). Their purpose, as Harvey's had been a century earlier, was to find the embryo at its earliest stages of development. Their results were inconclusive as far as proving preformation or epigenesis, since they were unable to find the embryo at all before seventeen days of development and could see no structures in it before nineteen days.[17]

In 1753, Haller returned to his native Bern in Switzerland to assume a life of civic affairs. There Haller began his own observations on incubated chicken eggs, which occupied him during the summer months of 1755, 1756, and 1757. Haller's notebook records of these observations ("Observationes anatomicae Bernenses") still exist in the Burgerbibliothek in Bern. But even more revealing of Haller's changing views on embryological matters during these years are the letters he wrote to his friend, the Swiss naturalist, Charles Bonnet.[18]

Figure 6. Charles Bonnet, 1720–93. (From the frontispiece to *Oeuvres d'histoires naturelle et de philosophie*, 1779–83; courtesy of the Syndics of Cambridge University Library)

Haller first mentioned the problem of generation in a letter written shortly after the instigation of their correspondence in 1754, where he remarked that he intended to begin working on the subject in the near future. To this Bonnet responded, "When you occupy yourself, Monsieur, with the great mystery of generation, the eyes of all physicians will be upon you" (Haller MSS, 28 September 1754). Asking in the same letter what Haller's views on the question were, Bonnet received a reply that indicates Haller was still thinking along epigenetic lines: "I have no system of generation. Without prejudice against evolution [preformation], I see more and more a simple, sticky material forming and shaping itself little by little. The egg of a sheep is for a long time all only a viscous fluid; it remains in this state 17 or 18 days. But it is necessary to repeat these experiments . . . " (Bonnet MSS, 14 October 1754).[19]

Haller's own experiments were not to begin until the following summer, and then they were on chickens rather than sheep. During the intervening winter months, Haller and Bonnet considered the question of the original fluid state of the embryo. As Haller remarked, still in relation to his experiments with Kuhlemann, "For the matter of generation, I believe that one part is clear. . . . A viscosity very surely successively congeals and becomes the membranes of the fetus. . . . [Yet] All is transparent and invisible" (Bonnet MSS, 26 November 1754). In his next letter, Haller hinted at how the transparency dilemma might be resolved: "I am strongly for your opinion, Monsieur, about the organic glue.[20] A salt dissolved in water retains invisible cubes. And there would be a jump if the small animal, becoming visible through the help of spirit of wine at the 18th day, already all formed, had been fluid the day before. It existed, without a doubt, approximately the same; but its transparency hid it from us" (Bonnet MSS, 4 January 1755).

As an epigenesist, Haller had used the original fluid state of the embryo as an argument against preformation and in favor of gradual development. Yet now he suggested that transparency may hide structures that actually exist in the viscous liquor. Although he did not develop the idea further at this point, it should be noted that Haller later made wholesale use of the argument that transparency hides existing structures in his campaign for preformation. It was also later to become an issue of controversy in his debate with Wolff.

Haller began his first series of observations on incubated chicken eggs in August 1755. Two major subjects occupied his attention: the formation of bone, which he was to investigate all along in his chick observations and to which he devoted a separate publication (1758b); and second, the formation of the heart and its relationship to the lungs. Haller had repeatedly used Malpighi's observation that the heart is originally only a simple tube that later develops chambers as an argument in favor of epigenesis. Yet in September, Haller wrote to Bonnet, "I have made many efforts to penetrate the mechanism of the heart. . . . I believe that I have caught a glimpse of that which does not change in its structure and that it is always equally formed of two auricles and two ventricles. . . . I spend two or three hours everyday reexamining this marvelous heart to give me an exact idea, which the vivacity of its movements renders difficult" (Bonnet MSS, 6 September 1755).

Haller's second series of observations, made during the late summer and early fall of 1756, cleared up the question of the development of the heart and lungs. As Haller reported to Bonnet, "For the heart, all the mystery that I investigated is reduced to a very simple thing. The lung, with its artery and its vein, is very small and invisible in the first times. Little by little it grows, and its artery which through its fineness and its transparent liquid was hidden from our eyes, becomes a second very considerable branch of the aorta from which two small branches lead to the lung, etc." (Bonnet MSS, 7 December 1756). Haller had been unable to reconcile the formation of the pulmonary circulation with Malpighi's notion of an originally tubular heart that curls up on itself to form the four-chambered heart, for the vessels of the lungs would have to break in at some point in this linear arrangement. Yet Haller solved this difficulty by proposing that, in early stages of development, the lungs and their vessels exist in an invisible state, connected to the heart as they will be in their later appearance. The heart itself develops from four originally existing but not clearly visible chambers. All simply become visible during development, with no fundamental changes taking place.

Haller's final series of observations were made during the spring and summer of 1757. It was during this period that Haller first announced his conversion to the theory of preformation. To Ignazio Somis, an Italian physician with whom

Haller frequently corresponded, he declared in July, "I find that all the changes of the egg relate much more to the system of evolution than to that of epigenesis" (Hintzsche 1965:40).[21] The final piece of evidence emerged when, having finished his observations on the heart, Haller began to turn his attention to the yolk and its relationship to the embryo. And in doing so, Haller discovered what he thought was a definitive proof of preexistence. This he announced to Bonnet on 1 September 1757:

> The membranes of the yolk are in a larger form the membranes of the intestines, but more developed. The yolk itself is only an expansion of the small intestine of the chick. It is from this that one can draw a very plausible reason to give the female the true beginnings of the fetus; for, in short, the yolk exists before the approach of the male; its membranes are continuous with those of the intestines of the chick; the chick thus appears to have existed before this approach. [Bonnet MSS]

Haller's "membrane-continuity" proof rested on the observation that, in later stages of development, the embryo's intestines are connected to the yolk sac in such a maner that the yolk sac seems to be simply a continuation of the intestines (see Figure 12, Chapter 3). Haller thought he had found the membranes of the yolk sac in the unfertilized egg; and he inferred from this, on logical grounds, that the embryo must exist at this time also, even though it cannot yet be seen. Consequently, the embryo exists in the female before fertilization by the male.

To Haller's announcement of his conversion and his new proof of preexistence, Bonnet, himself a staunch preformationist, jubilantly responded:

> It has been disputed for a century whether the embryo comes originally from the male or the female. . . . Monsieur de Haller opens an egg of the chicken, he examines it with eyes that have come to see, and he discovers that the yolk is an expansion of the small intestine of the chicken. The embryo thus existed in the female before the approach of the male. This discovery is assuredly one of the most important that anyone has made in the obscure matter of generation. [Haller MSS, 7 September 1757]

Later, after Haller had published these observations, Bonnet remarked simply, "Your chickens enchant me. I had not hoped that the secret of generation would be discovered so soon" (Haller MSS, 30 October 1758).

Haller read a report of his experiments before the Royal Society of Sciences of Göttingen in September and December

1757. These were published the following year in a two-volume work, *Sur la formation du coeur dans le poulet.* The basic tenet of Haller's new theory was that all essential structures of the embryo exist first in the female egg. When conception occurs, the heart is stimulated and, because of its inherent irritability, begins to beat.[22] The beating heart sends out fluids through the folded parts of the transparent embryo; the embryo's structures begin to solidify and to become more opaque; and the visible fetus emerges. "All of the parts of the animal body," Haller explains, "are born from fluid, apparently organized, which becomes mucous, and which acquires little by little more definite boundaries and a consistency that resists pressure" (1758a, 2:175). When fluids of increasing opacity, containing more and more viscous and solid particles, are pumped by the heart through the vessels, the parts of the embryo become increasingly visible as development proceeds. "The manner by which these same parts become visible from being invisible is of the most grand simplicity. It is the effect of growth, but even more the simple effect of opacity" (p. 176). As Haller summarizes:

> It appears very probable to me that the essential parts of the fetus exist formed at all times; not it is true in the way that they appear in the adult animal: they are arranged in such a fashion that certain prepared causes, hastening the growth of some of these parts, impeding that of others, changing positions, rendering visible organs that were formerly diaphanous, giving consistency to the fluidity and to the mucosity, form in the end an animal that is very different from the embryo, and yet in which there is no part that did not exist essentially in the embryo. It is thus that I explain development. [p. 186]

All organs and structures of the adult organism exist in the tiny, transparent embryo and are made apparent by mechanical causes.[23]

Haller's theory of preformation provided solutions to the difficulties he had previously raised when he had supported epigenesis. With regard to the initial fluid state of the embryo, Haller now saw no problem for preformation. Structures do exist in this initial viscous liquor (recall the dissolved salt analogy), even though they cannot be seen. Furthermore, Haller had found that, by using alcohol or vinegar, he could cause structures to become visible before their normal time of appearance (although he also found that there was a point prior to which this technique did not work). "After these observations," Haller remarks, "one should be on guard against the temptation to pronounce that such and such a part of an animal is newly produced and that it has not existed before. It could have been too small; it could have only been transparent" (1758a, 2:177). Appearances should not deceive us into thinking that gradual development is actually occurring.

Haller was also now able to solve the problem of changing structures in the embryo. As an epigenesist, he had pointed to the alterations one observes in major organs, especially the heart, as supporting gradual development. Yet through his own observations, Haller was able to discount Malpighi's description of the formation of the heart, for Haller believed he could show that all of the changes one observes are only apparent. This conclusion he broadened to apply to the whole embryo, and he postulated that differential rates of growth and increasing opacity could account for seemingly drastic alterations in structure. As Haller later remarked, "If I myself, more than twenty years ago, before my repeated observations on eggs and on pregnant quadrupeds, once used the argument that the fetus is very different from the fully grown animal and therefore that there are not found in the rudiments of the new animal those parts that are in the same, now fully grown animal, I recognize manifestly now that all the very same arguments that I brought forth against evolution [preformation] were in support of it" (1757–66, 8[1766]:148–49). The changes one observes in the embryo's structure are not evidence for gradual formation but rather for development from preexisting parts.

Haller's other two previous objections to preformation – the problems of hybrid organisms and regeneration – were now accounted for as well. Hybrids resemble the male parent, Haller proposed, because the male semen can cause greater growth by speeding up the movement of fluids in certain parts of the embryo (1758a, 2:189–91). Hence, for instance, the long ears of the mule and other unusual structures found in hybrid organisms. Yet exactly how the male semen causes this increased movement of fluids in specific embryonic parts is never clearly articulated by Haller.

Haller explained regeneration in a similar ad hoc manner. In his *Elementa physiologiae*, he adopted the position, argued by Bonnet among others, that organisms such as the polyp and the crayfish can regenerate lost parts because they contain preexistent germs (1757–66, 8[1766]:170–71). When the organism is injured, these germs can be called into play. Never expanding on the nature of these preexistent "head germs" and "tail germs," Haller argued in direct contradiction to his earlier views that regeneration is not a case of gradual, epigenetic development. Some organisms possess preexistent germs and are thereby capable of forming a lost part through the development of these germs. Neither Haller's explanation for regeneration nor that for hybrids was very satisfactory, yet he no longer saw these two areas of objection as fatal to the theory of preformation.

Haller's new explanation of embryological development solved entirely the dilemma concerning the operation of forces that had arisen in his critiques of Buffon's theory. As we can recall, in dealing with Buffon's ideas on generation, Haller came face to face with the problem of how development is organized and directed. Because of his own beliefs, Haller could not accept either self-directed material forces or immaterial, vital forces. He was thus left with forces of a material but simple nature, operating mechanically, in the same way that the force responsible for irritability operates. Yet these forces could not, on their own, produce development without being directed in their operations by God. It was this key issue of guidance, which Haller had left unexplained in his epigenetic theory, that reopened the problem of generation for him and prompted first the experiments on sheep with Kuhlemann and then the observations on incubated chicken eggs.

Haller's theory of ovist preformation provided a clear mechanism for the operation of forces in development and for God's guidance. All forces that are part of the developmental process are simple and mechanical in nature; they possess no self-directing abilities whatsoever. Their utilization and guidance is based ultimately on God who, at the Creation, constructed all of the preformed organs of each future embryo and, in addition, organized in these structures all of the forces that would be active in the embryo's later development. As Haller explained in his *Elementa physiologiae*:

> If the first rudiment of the fetus is in the mother, if it has been built in the egg, and has been completed to such a point that it needs only to receive nourishment to grow from this, the greatest difficulty in building this most artistic structure from brute matter is solved. In this hypothesis, the Creator himself, for whom nothing is difficult,

has built this structure: He has arranged at one time, or at least before the male force [of fecundation] approaches, the brute matter according to foreseen ends and according to a model preformed by his Wisdom. [1757–66, 8(1766): 143]

When the tiny embryo is stimulated by the male semen, the heart, because of its inherent irritability, begins to beat, initiating the entire preprogrammed sequence of development. Everything takes place by natural, material forces, yet all is ultimately under the guidance of God.

Haller's observations on incubated chicken eggs provided two major sources of evidence for his preformation theory. First, he thought he could now show that the heart does not develop as Malpighi believed, but rather in a manner entirely consistent with preformation. Second, Haller felt that he could actually prove that the embryo exists in the unfertilized egg through his membrane-continuity proof.[24] Haller's observations on chicken eggs combined to provide the catalyst for his conversion, for they enabled him to work out a viable theory of embryological development based on preformation. Yet it is also clear that the need to develop such a theory, one that would be consistent with his fundamental religious and mechanistic beliefs, prompted Haller to begin these observations and in this sense prepared him for the discoveries that he then made. The period between the late 1740s and the observations on chicken eggs in 1755 to 1757 witnessed Haller's realization of what the implications and problems of epigenesis were, and his growing uneasiness with how these difficulties could be resolved in a manner consistent with his own philosophical beliefs.[25] These factors combined to set the stage for Haller's observations and, through them, to produce his conversion to preformation.

3
The embryological debate

Within a year of the publication of Haller's *Sur la formation du coeur dans le poulet* there appeared a work by a young Berlin physician, Caspar Friedrich Wolff, that strongly supported epigenesis. Wolff sent a copy of his *Theoria generationis*, which had been his doctoral dissertation at the University of Halle, to Haller, in the hope that Haller would give up his new defense of preformation. Little did Wolff suspect that his actions were to provoke a ten-year controversy with Haller, who, rather than being convinced by Wolff's work, became preformation's staunchest and most rigid defender. In this and the next chapter, I shall analyze the debate that ensued between Haller and Wolff. This chapter will concentrate on the embryological aspects of the debate, while Chapter 4 will focus on the underlying philosophical themes of controversy. By way of introduction, let me present here an overview of the major events of the debate. (The reader is referred also to Appendix A, "Chronology of the Haller–Wolff Debate.")

The controversy between Haller and Wolff was manifested in a series of publications and in several letters that passed between the two.[1] Additionally, Haller published reviews of each of Wolff's works in the *Göttingische Anzeigen von gelehrten Sachen*. It was in response to the first of these reviews, in which Haller criticized Wolff's dissertation, that Wolff published his second book, *Theorie von der Generation*, in 1764. Not a translation of the *Theoria generationis*, as some historians have surmised, the *Theorie von der Generation* contains rather a restatement of Wolff's theory and a polemical attack on the preformationist views of both Haller and Bonnet.

Haller published a brief review of this work, but his major response appeared in the eighth volume (1766) of his *Elementa physiologiae corporis humani*, which also contains a more detailed account of his own theory of preformation. During the next two years, there appeared Wolff's "De formatione intestinorum" in the journal of the St. Petersburg Academy of Sciences (1766–67, 1768). This work was devoted to demonstrating the epigenetic development of the intestinal tract and was perhaps the most influential of all of Wolff's publications, at least with regard to his nineteenth-century successors in the field of embryology.

Wolff moved from Berlin to St. Petersburg in 1767. Thereafter, little communication took place between Haller and Wolff, and their debate effectively came to an end. Haller died in 1777, before the publication of Wolff's last major embryological work, *Von der eigenthümlichen und wesentlichen Kraft* (1789), which contains both a further explanation of Wolff's theory and some reflections on Wolff's debate with Haller. Wolff's own death came in 1794, before he was able to complete his last major work, on the subject of monstrous births.

Following an analysis of Wolff's theory of embryological development, this chapter will be organized around the principal themes of controversy that emerged in the Haller–Wolff debate. These embryological points of contention fall into three major areas: the formation of blood vessels in the area vasculosa, the development of the heart, and Haller's membrane-continuity proof of preformation. I shall attempt both to describe the exchanges between Haller and Wolff on these issues and to discuss their controversies in more modern embryological terms. Finally, Wolff's work on the formation of the intestines will be presented, followed by some concluding remarks on embryological observation and on the debate as a whole.

WOLFF'S THEORY OF EPIGENESIS

Prior to studying at the University of Halle, Wolff attended the Collegium Medico-Chirurgicum in Berlin, moving to Halle in 1755 at the age of twenty-one. Wolff stayed at the university for four years, afterward becoming a military physician in Breslau during the Seven Years War. Upon returning to Berlin in 1763, Wolff requested permission from the Collegium Medico-Chirurgicum to offer public lectures on physiology. After considerable controversy, sparked perhaps by Wolff's uncommon embryological theory, he was allowed to give private instruction on logic, physiology, and anatomy. When two positions became vacant at the Collegium in 1764, Wolff applied for a professorship but was unsuccessful. Finally, in 1767, Wolff received an invitation, on the recommendation of Leonard Euler, to be-

Figure 7. The only known portrait of Caspar Friedrich Wolff, a silhouette done in 1784. (From Gaissinovitch, *K. F. Vol'f i uchenie o razvitii organizmov*, 1961)

come Professor of Anatomy and Physiology at the St. Petersburg Academy of Sciences. He moved to St. Petersburg the same year, remaining there until his death.[2]

Wolff's first publication was his doctoral dissertation, which he defended at the University of Halle in 1759. In this treatise, Wolff proposed a model for development in plants and animals based on two factors: the ability of plant and animal fluids to solidify, and a force, which he named the *vis essentialis* ("essential force"). In Wolff's system, the origins and subsequent formation of the embryonic plant or animal are produced through the secretion of fluids that then solidify into structures. Beginning with the seed in plants and the yolk (or placenta) in animals, this secretion–solidification process proceeds in serial order, with each part secreting the next after its own formation. As each part begins to solidify, it becomes "organized," acquiring vessels and vesicles that are produced by the movement of fluids into the new part.[3] (Vesicles are formed when fluids move into a homogeneous part and form stationary pockets, whereas vessels are formed by the movement of fluids through a part.)

Wolff postulated that in plants there exists a "vegetation point" at the tip of each growing stalk, from which leaves, blossoms, or fruit develop through secretion and solidification (1759:24–30, §§ 43–53). Since the particular type of structure that is produced at a given time is dependent upon the amount of nourishing liquid that reaches the vegetation point, Wolff claimed that all of these structures are essentially "modified leaves."[4] At the vegetation point, the nourishing liquid generates the first rudiments of leaves by secretion of fluid that then solidifies. This is why, Wolff maintains, one can peel off leaves or buds at the tips of growing stalks and find smaller rudimentary ones underneath. This phenomenon, that new leaves could be found folded up in miniature inside older ones, had frequently been cited as a classic case of preformation. Yet, Wolff argues, this is actually an example of epigenetic development from the vegetation point.

In plants, Wolff asserts, vegetation (development) occurs throughout their lives. But in animals, this process takes place only during the early stages of each organism's life. The first structure to solidify in higher animals, according to Wolff, is the spinal column (see Figure 8; Wolff's "spinal column" with its "vertebrae" is actually the somites). Wolff points also to the formation of the kidneys as a secretion from the spinal column,

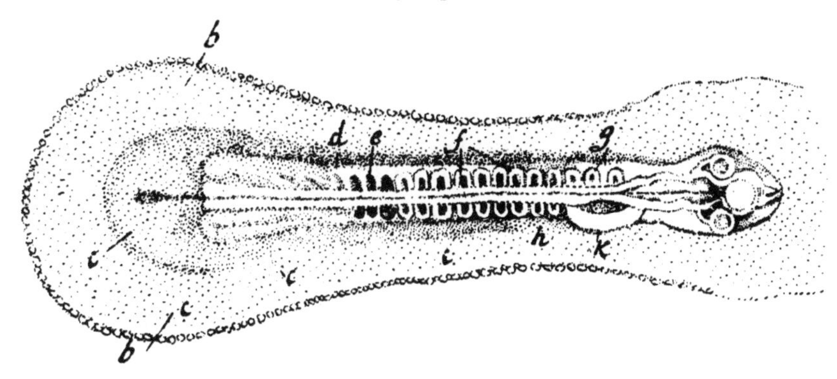

Figure 8. Wolff's illustration of a chick embryo at 36 hours: *b*, amniotic cavity; *c*, material secreted from the spinal column, beginning to condense into the limbs; *d, e, f, g*, rudimentary vertebrae; *h*, spinal cord; *k*, heart. (From *Theoria generationis*, 1759; courtesy of the Francis A. Countway Library of Medicine, Harvard Medical School)

first in the form of an unorganized mass that then splits into two halves, each of which gradually acquires vessels and vesicles. Wolff was actually observing the temporary embryonic kidneys, now called "Wolffian bodies," unaware that these are not the kidneys that exist in the mature chick.

All of these developmental processes are produced by the essential force that Wolff felt could be postulated on the basis of observational evidence. In plants, he points out, fluids are constantly being absorbed from the soil, distributed throughout the plant, and evaporated from the leaves. What else but a force could cause this movement? "It must be assumed [to exist]," Wolff contends, "if plants and their nourishing liquids are granted, which has been confirmed through experience: This will be sufficient for the present purpose, and it will be called by me the essential force of plants" (1759:13, § 4). In chicken eggs, Wolff observes, the embryo increases in size from the first moments of incubation, before it possesses a visible heart. Material must be entering the embryo from the yolk, and "it follows, that the nourishing particles pass from the egg into the embryo, and that there exists a force, through which this is accomplished. . . . I will call this similarly the essential force" (p. 73, § 168).

Through his secretion–solidification model and the essential force, Wolff believed he could offer a complete explanation for embryological development. "The essential force," he concludes, "along with the ability of nutrient fluid to solidify, constitutes the sufficient principle of all vegetation [development] both in plants and in animals" (p. 115, § 242). There is no need to assume the existence of preformed parts in the embryo; everything is formed gradually, through the secretion and solidification of fluids under the guidance of the essential force.

Wolff's theory was markedly different from Haller's. Well-argued and based on observational evidence, Wolff's epigenetic explanation of development understandably presented a clear challenge to Haller, who had not long before rejected his own epigenetic beliefs in favor of preformation. Let me turn now to the major issues of controversy that emerged in their ensuing debate.

THE EMBRYOLOGICAL DEBATE

The area vasculosa

Haller and Wolff devoted more time to debating the formation of blood vessels in the area vasculosa than to any other single issue. The gradual appearance of this vascular network is indeed a striking event in chick development, one noticed by the earliest investigators. These blood vessels, known today as the vitelline circulatory arc, serve to bring nourishment from the yolk to the embryo. They form in the area vasculosa, a region of the blastoderm surrounding the developing embryo, and eventually cover a sizable portion of the yolk's surface, bounded on the outer edge by a circular blood vessel called the sinus terminalis.

In the developing chick embryo, three separate circulatory arcs are formed, serving different functions. The embryonic circulation, which develops within the body of the embryo, and the vitelline circulatory arc, which develops in the area vasculosa, form approximately contemporaneously. The third circulatory arc, the allantoic, is formed later and carries blood to the allantois, a saclike structure where waste products are stored and where the blood is oxygenated through a region of the allantoic blood vessel network (the chorioallantois) that lies just under the shell. In mammals, which have only a small vitelline circulatory arc (and no appreciable amount of yolk), the allantoic circulatory arc is commonly called the umbilical circulation. Through its connection with the placenta, the umbilical circulation carries on the functions of nourishment, waste disposal, and oxygenation, functions that are separated in the chick between the vitelline and the allantoic circulatory arcs (see Adelmann 1966, 3:1104–5; Patten 1971).

The development of the vitelline vessels in the area vasculosa exhibits striking changes as incubation continues. In early stages of development, the area vasculosa presents a mottled appearance, due to the formation of blood islands by the end of the first day. These begin to form at the periphery of the area and later extend into regions closer to the embryo. After about 33 hours of incubation, the area vasculosa begins to look more netlike because of the extension and joining together of blood islands to form blood vessel channels. Within the newly forming vessels, primitive blood corpuscles begin to appear, and the rudiments of the sinus terminalis become visible. This network continues to develop, progressing inward toward the embryo to join with the major vein and artery (the omphalomesenteric vessels) that connect the embryo's heart to the vitelline circulatory arc through the yolk stalk. By about 40 hours of incubation, the vitelline blood channels are completed, and the heart develops a sufficiently strong heartbeat for circulation to commence.

As mentioned earlier, the formation of blood vessels in the area vasculosa is one of the most visible aspects of development in the early days of incubation. Noted by Aristotle, Fabricius, Harvey, and others, the stages of development of the area vasculosa were first most completely described by Malpighi.[5] Beginning with an embryo of 12 hours' incubation, Malpighi described the appearance of what he called the umbilical vessels, offering a remarkably accurate account in gross detail (see Figures 1 and 9). Malpighi never clearly separated the vitelline from the allantoic circulation, nor did he fully understand the analogy of the avian umbilical vessels with the mammalian; these confusions were clarified to a great degree by Haller's observations. Between Malpighi and Haller, little new became known about the area vasculosa, and it was really the work of both Haller and Wolff that represented the first major advance over Malpighi's observations.

Before the controversy between Haller and Wolff, the area vasculosa did not figure as a major issue among either preformationists or epigenesists. Harvey believed that the umbilical vessels are formed gradually, yet the area vasculosa never became a principal example of epigenetic development for him (1651, 1847 trans.: 235, 238, 392–97). Malpighi expressed the belief that the umbilical vessels preexist, only gradually coming into view as they are filled with fluids and finally with red blood. Yet one must be careful to note, as Adelmann has pointed out (1966, 2:885–86), that when Malpighi maintained that a struc-

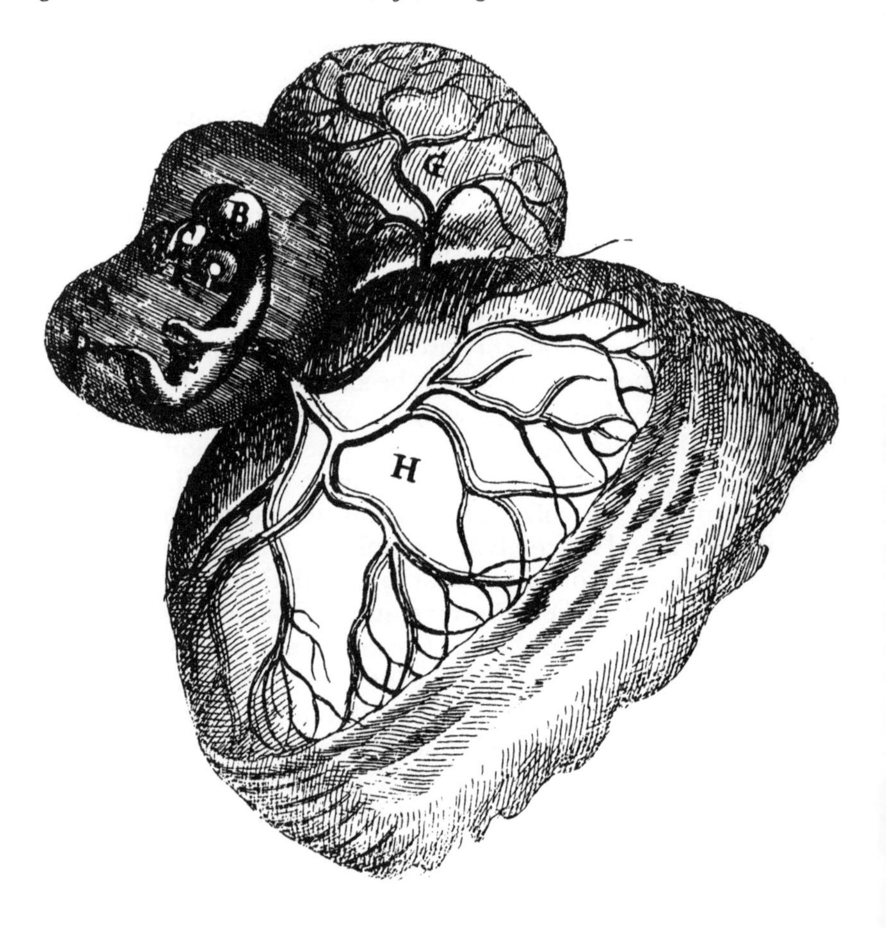

Figure 9. Malpighi's illustration of a chick embryo in relation to the yolk, and of the vitelline circulation in the area vasculosa (at 6 days). In modern terminology, *A*, the amnion containing the embryo; *G*, the allantois; *H*, the area vasculosa and vitelline vessels. (From *Opera omnia*, 1686; reproduced by permission of the Houghton Library, Harvard University)

ture preexists, he did not mean that it exists before fertilization but only that the particular structure exists before incubation. As Malpighi remarked in his *Dissertatio epistolica de formatione pulli in ovo* (1673), "while . . . we are studying attentively the genesis of animals from the egg, lo! in the egg itself we behold the animal already almost formed, and our labor is thus rendered fruitless. For, being unable to detect the first origins, we are forced to await the manifestation of the parts as they successively come to view" (trans. Adelmann 1966, 2:935–37). Although formed in essential aspects before incubation, the chick embryo presents successively different appearances, according to Malpighi, both because of preexistent structures coming to view and because of substantial changes in these structures during development. In the area vasculosa, Malpighi claimed, it is principally a matter of increasing visibility that is responsible for the successive appearance of these blood vessels; for the "complete plexuses of umbilical vessels pre-exist in the cicatricula and as a result of the turgescence and motion of the ichor entering them become visible, extend their territories, and form trunks" (1675, trans. Adelmann 1966, 2:997).

Thus, although Malpighi believed that the blood vessels of the area vasculosa preexist and only gradually become visible, he did not use this example to support a preformationist position. It was Haller who first made a strongly preformationist case for the gradual appearance of the vitelline vessels, and it was Wolff's equally forceful proponency of an epigenetic account of their development that brought the area vasculosa to the fore as a principal focal point in the preformation–epigenesis struggle.

Haller described the development of blood vessels in the area vasculosa (called by him variously the *membrane ombilicale* and the *figure veineuse*) in his *Sur la formation du coeur dans le poulet*. "At the commencement of the chick," he writes, "it is a soft pulp: netlike traces are born in this pulp by the force of the heart; they begin to be points; they become lines; the lines become colored and are the arteries and veins, divided at very small angles. These angles enlarge; white areas form between these vessels; they dilate in time, just as the spaces comprised between the veins of leaves dilate" (1758a, 2:173–74). As a description of what one observes, Haller's account squares fairly well with the appearance of the area vasculosa at successive stages of development. Yet, he continues, "if one goes backwards in considering the successive changes in the umbilical membrane, one will be easily convinced that it has always existed with its vessels, that it has been folded upon itself, that the impulsion of blood has elongated the arteries, or unfolded these folds, and that it has moved the vessels away from one another and has given the membrane its width, its length, its white areas, even its solidity" (p. 174). The pumping of fluids by the heart through the folded vessels of the area vasculosa produces the appearance of gradual development that one observes. As Haller concludes, "I believe that this example is very instructive, and suitable for demonstrating the nuances by which a soft and semifluid material can pass to a state entirely different from its primordial state, by the simplest of evolutions" (p. 174).

Thus, for Haller, the area vasculosa became a prime example of his mechanism for development from preformed parts. Originally a semifluid, transparent material, the area vasculosa becomes transformed into a complex web of blood vessels simply by the pumping through transparent vessels of fluids that become more and more opaque and gradually colored. Through this same sequence of events, solidification, expansion of parts, and increasing opacity are produced in all of the embryo's preexisting primordial structures. The area vasculosa was for Haller the most striking case of this developmental process.

In Wolff's *Theoria generationis*, published a year later, the area vasculosa was again used as a principal example, but this time for epigenesis. In early stages of development, the embryo is surrounded, Wolff maintains, by a series of "rings" that form in the area vasculosa (called by Wolff the *area umbilicalis*). These rings, illustrated in Figure 10 (*top*), are formed, according to Wolff, by the movement of fluids from the yolk to the embryo. During the first moments of development, the yolk substance begins to dissolve under the warmth of incubation and to move toward the embryo, guided by the essential force. The rings are formed, Wolff says, from nonnutritive deposits that are left behind as the dissolved yolk moves to the embryo. Wolff's description of the movement of yolk substance inwards toward the site of the developing embryo is reminiscent of Malpighi's discussions of the movement of colliquament (embryonic fluid) through the umbilical area toward the fetus. Malpighi had stated that, in the earliest stages of development, the area vasculosa in a chick egg seems to undergo a melting process. It "was not everywhere a solid structure," Malpighi noted, "but, like a hill washed into and flooded by gushing springs, was being gradually dissolved by colliquament invading it from outer rivulets" (1673, trans. Adelmann 1966, 2:947). He even speculated that these rivulets might be the primordial umbilical vessels, which later become visible as colored fluid moves through them, yet he returned again to the idea that these blood vessels are preformed (Adelmann 1966, 2:949). Even the term "colliquament" gives rise to comparisons with Wolff's later observations, as it was coined by Harvey to denote the fluid out of which the embryo is produced, formed from the colliquescence, or melting, of the albumen and yolk.

Yet Wolff's model, similar though it may be on gross observational terms to Malpighi's, was unique to Wolff's epigenetic system. The movement of dissolved yolk substance into the embryo provided evidence, Wolff believed, for the existence of the *vis essentialis*. Because a heart cannot be observed in the early hours of incubation, it could not be responsible for the observed movement of fluids. Therefore, "it follows that the nourishing particles pass from the egg into the embryo, and that there exists a force, through which this is accomplished" (1759:73, § 168). It is the essential force that is responsible for the movement of these fluids and thereby for the formation of the blood vessels of the area vasculosa.

After the initial formation of rings in the area vasculosa, one observes, according to Wolff, the breaking up of these rings into "islands" surrounded by dissolved material. Gradually these islands become smaller and smaller, until a network of connected channels is formed.[6] These channels, Wolff claims, are the initial rudiments of the blood vessels, obtaining vessel walls only when the material immediately surrounding them begins to thicken. (This occurs after the blood is propelled through the vessels by the newly formed heart.) As the network of channels increases in complexity, major branches appear, and the completed vitelline circulation emerges (see Figure 10, *bottom*).

Wolff sent Haller a copy of his dissertation on 23 December 1759.[7] In his accompanying letter, Wolff indicates that he hopes that his work might sway Haller's recent conversion to preformation and convince him again of epigenesis. "What then prevents me," Wolff writes, "having completed my little work, from submitting it now to your most penetrating judgment as reasons fighting for the other side?" (Haller 1773–75, 4:269). Little could Wolff have guessed the reaction his treatise was to provoke from Haller.

Less than a year later, Haller published an anonymous review of Wolff's *Theoria generationis* in the *Göttingische Anzeigen von gelehrten Sachen*.[8] Opening with the words, "We have not read in a long time a work as important as that of Carl Friedrich Wolff" (1760:1226), Haller's review presents a fair summary of Wolff's theory of embryological development. Yet Haller charges that Wolff's entire model rests upon an unwarranted assumption: that if one does not observe a structure, then one can conclude that it does not exist. "But," Haller responds, "whoever has practiced much himself with the magnifying glass will have

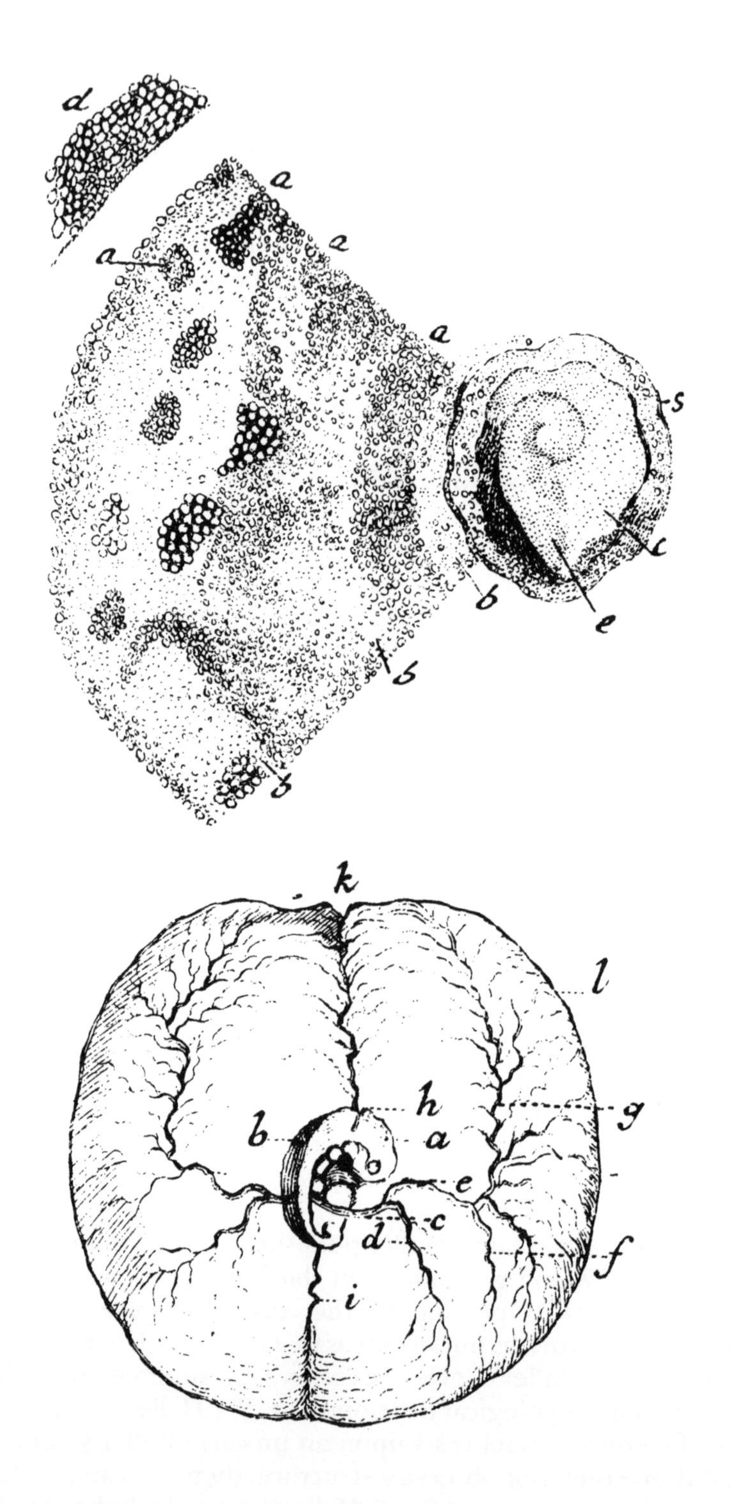

Figure 10. Wolff's illustrations of chick embryos and the area vasculosa. *Top,* after 28 hours of incubation. The coarser material (*a*) is broken up by fluid material (*b*),

informed himself particularly in the mesenteric veins of frogs that indeed strong colors make parts visible and transparency makes them invisible; and in the grown animal, whose veins certainly have in other cases visible membranes, very often the blood corpuscles are visible, without one being able to recognize the membranes of the veins" (p. 1229). There are clear cases, Haller argues, such as the mesenteric veins of the frog where transparency does indeed hide the existence of blood vessel walls but where we know they exist on the basis of other evidence. "This remark is here all the more important," Haller continues, "because Herr Wolff believes he has proved, with great reason and with the testimony of other observers of nature, that in the so-called *Area umbilicali* in the chick roads are drawn, which gradually become perfected into vessels." Yet, Haller concludes, "The appearance is correct; only the doubt remains whether the transparent roads through the granular material exist truly as mere roads without membranes, and this is not so easy to make out" (p. 1229).

Wolff responded to Haller's review of his dissertation in his second major publication, *Theorie von der Generation* (1764). He even reprinted the anonymous review under Haller's authorship, much to Haller's chagrin, as is made evident in the ensuing correspondence between Haller and Wolff (see Appendix B) and between Haller and Charles Bonnet. Wolff also attacked Bonnet rather viciously in this work for uncritically supporting a preformationist position closely allied with, and based largely on, Haller's (see Bonnet 1762).

Wolff answered Haller's charge that his system is based on the principle "what one does not see, is not there" by arguing that his theory of epigenesis does not in fact rest upon this principle at all. Rather, Wolff claims, where he had argued that a particular part does not exist, he had done so on the basis of independent evidence.

This is especially the case, according to Wolff, with regard to the formation of blood vessels in the area vasculosa. One can actually *observe* the gradual formation of these vessels; one does not simply assume that they do not exist before they are visible.

Figure 10 (*cont.*).
forming rings and islands; *e*, the embryo seen through the amnion. *Bottom*, after 3½ days of incubation: *a*, the embryo; *c, d, e, f, g, h, i*, major blood vessels of the area vasculosa; *l*, vena terminalis (sinus terminalis). (*Top*, from *Theoria generationis*, 1759; courtesy of the Francis A. Countway Library of Medicine, Harvard Medical School. *Bottom*, from "De formatione intestinorum," 1766–67)

Repeating his description of the formation of islands and channels in the area vasculosa, Wolff asks, "Can I therefore say, that in the circumstances, in which the area is . . . [in early stages of incubation], vessels are already contained in it? Rather, must I not say, that they are not only engaged in building but that only the very first stage in the building of vessels has begun? and that consequently true vessels are not yet there?" (1764:87). There is no basis, Wolff asserts, for maintaining that blood vessels preexist in the area vasculosa when one can actually observe their gradual formation. Consequently, Wolff continues, "do I assert . . . that there are no vessels because I do not see them, and I make this conclusion on the principle what I do not see, is not there? Rather do I not conclude these truths from that which I really see, because I see that the vessels for a long time are not finished, that they are scarcely first begun and actually not yet vessels?" (p. 88). True vessels do not exist in the early stages of development. This one can conclude not by inference but because one can actually see them form.

Wolff charges that Haller's account of the unfolding of preexisting vessels is contrary to observation in two ways. First, Wolff claims that as the material surrounding the channels increases in density one can observe the formation of vessel walls. Consequently, these vessel walls could not have preexisted in a transparent state, as Haller claimed. Second, Wolff asserts, if preexistent blood vessels become visible only when the heart pumps colored fluid through them, then one should observe the islands of material in the area vasculosa separate from one another in a regular, orderly manner, "as if they were cut with a knife from one another" (p. 94). But what one actually observes is large islands separating irregularly from one another into smaller and smaller pieces. "You do not see the slightest trace, but rather the opposite," Wolff asserts, "of an actual expansion and extension by the penetration of blood pumped from the heart through such vessels, which are already there, and were only folded together. And that is thus the second contradiction with nature the ideas of Herr von Haller make" (p. 95).

Furthermore, Wolff argues in an appendix to his book based on new observations on incubated chick eggs, if one looks at the area vasculosa at an even earlier point in time, all you see is cavities in the granular material, with channels not yet formed. These do not communicate with one another and are more numerous at the periphery of the area than near the embryo's location. This could not possibly be explained by Haller's account, for "that they are vessels or even only passages that . . . were expanded and thereby made visible through the force of the heart by means of the liquids pressed into them is impossible, because these cavities do not communicate with one another, and no passages toward the heart are present through which the liquids would have had to come" (p. 262). The pumping action of the heart could not possibly have been responsible for the appearance of the area vasculosa at this early stage in development.

Wolff sent Haller a copy of his new book and wrote that for the book he had "undertaken new experiments on incubated eggs, so that I might render the theory of generation more firm" (Haller 1773–75, 5:210; letter of 20 December 1764). Wolff calls special attention to his observations on the area vasculosa, enclosing a series of illustrations (see Figure 20, Appendix B; see also Belloni 1971). Two of Wolff's drawings depict the secretion and solidification of the limbs from the spinal column. The remaining two illustrate the formation of blood vessels in the area vasculosa and the islands and channels produced by the movement of fluids during early stages of development. Wolff reiterates his argument that, because these islands and cavities begin to form first in the outer portions of the area vasculosa rather than in the region nearest to the embryo, they could not be produced by the pumping of the heart. As he concludes to Haller, "The integrity of the area in the places closest to the amnion [and the embryo], which cannot be divided even by a needle without evident violent laceration, prevents you, illustrious one, from being able to regard these fissures as perfected vessels that have only been dilated by the force of the heart and by blood propelled from the embryo" (Haller 1773–75, 5:211–12).

Haller received Wolff's letter and the accompanying book sometime in February 1765, as he reports in a letter to Bonnet. By March, Haller decided to resume his observations on incubated eggs.[9] Writing to Bonnet he explains, ". . . this M. Wolff forces me to begin again experiments on eggs. It is a matter of recognizing if the *traces* [in the area vasculosa] are vessels or if these are globules without partitions. Is he not amazed that the globules are formed from a circular vein and can come back to the heart? The globules diffused in the mesentery of a frog have certainly not undertaken a similar task" (Bonnet MSS, 10 March 1765). The blood corpuscles contained in the mesenteric veins of the frog move in an orderly way because they are contained in vessels with transparent walls. How, Haller asks, could vessels formed in the outer regions of the area vasculosa find their way *back* to the embryo's heart?

Yet Bonnet was unwilling to give any credence to Wolff's views. As he remarked to Haller, "I have built on your facts as on solid rock. Has Wolff made a cut in this rock? These vessels that fashion themselves, according to him by an aggregation of molecules, would persuade me only of the prejudice of the observer" (Haller MSS, 19 March 1765). Bonnet had indeed built much of his own theory on Haller's observations (see Bonnet 1762), and the appearance of Wolff's new observations was clearly an unwelcome event.

Haller was inclined to take Wolff's explanation of blood vessel formation in the area vasculosa more seriously. He explains in his next letter to Bonnet, "I have but one observation that relates to M. Wolff: it is enough to see again that the *traces* and *points* are only the folded vessels, and that they appear red only in the ends [that are] the fuller. All the rest is no longer important to me; but this single phenomenon establishes the system of M. Wolff or destroys it" (Bonnet MSS, 4 April 1765). Haller's sense of urgency with regard to Wolff is underscored in a further letter, where he writes, "I am reading Wolff: it is absolutely necessary to review eggs with the microscope to satisfy oneself on the existence of the membranes of the vessels and on the preexistence of the heart, which this physician claims to have seen born" (Bonnet MSS, 11 April 1765).

Alarmed by Haller's letters, Bonnet responds, "What! You are called again to observe *eggs*! Wolff has elevated in your spirit doubts powerful enough to persuade you of the necessity of resorting to a new examination! He has seen a *heart* born that did not exist before! And this *birth* is real! It is not an appearance! There is a *generation* properly called! Nothing is *preformed*! In truth, it would be necessary to knock me over the head with the proofs for me to admit it" (Haller MSS, 19 April 1765). But, Haller reassures Bonnet, "I will make experiments on eggs, and I do not doubt that he has been mistaken" (Bonnet MSS, 23 April 1765). Haller continues in the same letter, "There appear in the *figure veineuse* [area vasculosa] traces and points, following which there are complete and continuous vessels. I take these traces for the true vessels, of which a part is transparent. M. Wolff takes them for *roads* through which the grains of the venous substance, put in solution, are transported to the fetus." Later, Haller writes again, "Wolff is not without genius: I find that he has been mistaken on several points: I hope that he will have been on others; but it is necessary to read the book of nature to refute him" (Bonnet MSS, 19 May 1765).

Haller published a review of Wolff's *Theorie von der Generation* in June in the *Göttingische Anzeigen von gelehrten Sachen* (1765a). Much less critical than his review of Wolff's dissertation, this review is principally summary. It was not until he began a new set of observations on incubated chicken eggs in July that Haller developed new ammunition to combat Wolff. By mid-August, Haller was able to report success to Bonnet. "This is a month, my illustrious colleague, that I pass in experiments on eggs. They are sufficiently consistent with the preceding ones" (Bonnet MSS, 25 August 1765). Haller proceeds to present some observations on the connection between the embryo's intestines and the yolk, which I shall treat at a later point in my discussion. Then he turns to the area vasculosa:

Are the vessels of the *umbilical area* (a name that it is necessary to correct) or the *figure veineuse* vessels or passages that the nourishing liquid forms in a pulpous material?

I have tried two experiments to decide this question. I have planted the point of a very fine lancet in the *roads* that are still yellow or pale and without redness; I have planted it there when they have become red and appear as traces and spots. The *roads* are not deranged by the scalpel; the blood is not spilled; the road has followed the point of the scalpel to the right or to the left, and it is reestablished when one puts it back in freedom.

I have explained favorably this phenomenon. The point, too blunt to pierce the membrane of a very fine vessel, has only drawn it along without opening it. If it were a road without a membrane, this blood would be spilled, the road would be enlarged and would become irregular, etc.

Because the blood vessels Haller attempted to pierce with the scalpel did not rupture but simply moved from side to side, Haller concluded that the vessels could not be "mere roads without membranes" as Wolff believed. Consequently, true preexistent vessels were demonstrated in the area vasculosa.

Haller's second experiment was as follows: "The other experiment was made with vinegar, which turns blood dark. If you pour it on the roads, it does not alter the color of the blood for a long time afterward; and the acid certainly penetrates perfected vessels, it darkens the blood little by little in the veins, but it does not do it all at once as it does when it touches the blood immediately." If the blood vessels were channels with no walls, then the vinegar should reach the blood all at once, and an immediate change in color should take place. But, Haller argues, a gradual color change indicates that the vessels have membranes because the vinegar reaches the blood more slowly. This is what one observes when one places vinegar on mature vessels. Consequently, true vessels must exist in the area vasculosa. Haller concludes triumphantly to Bonnet, "Thus, my illustrious colleague, this will make the foundation of a new supplement on eggs; I believe that M. Wolff will be quieted down." To which Bonnet responds, "Your new observations on the chicken, my illustrious friend, have given me indescribable pleasure . . ." (Haller MSS, 17 September 1765).

With the results of these experiments in hand, Haller communicated them to Wolff. As he reported to Bonnet, "I have written to M. Wolff; I do not know if he will capitulate" (Bonnet MSS, 28 September 1765). This idea certainly appealed to Bonnet, as he remarked in reply, "I am very impatient to know the part that M. Wolff takes after having read and meditated on your critique. It pleases me to think that he will capitulate in true philosophy, and this victory that you obtain over him will be as glorious as you yourself" (Haller MSS, 8 November 1765).

Haller's next published work on embryology was the eighth volume of his *Elementa physiologiae*, which appeared in 1766. Responding to Wolff was clearly on Haller's mind, as he devoted more pages to discussing Wolff than to any other single figure (with Buffon a close runner-up). The text of this volume must have been written before Haller's summer egg observations, because these are discussed only in a section of Addenda at the end of the volume. In the text, Haller presents several other arguments against Wolff's views on the formation of the blood vessels in the area vasculosa.

First, Haller argues that most mature blood vessels have muscular fibers in their vessel walls and nerves wrapped around them. But no one, he maintains, has ever found that muscle or nerve fiber can be newly produced in the body. Thus, how can a vessel be formed initially by the passage of fluids through material? The vessels must exist, along with the associated muscles and nerves, before development. Haller then counters Wolff's argument that one can actually observe the gradual formation of the blood vessels by claiming once again that, before red blood penetrates them, the vessels, and the pale fluid they initially contain, are too transparent to be visible. "Thus," Haller maintains, "... the vessels of the yolk that form the *figura venosa* have been present for a long time even before the blood with its globules came to it. It is only the pale fluid contained in them that has hidden them" (1757–66, 8[1766]:276). Wolff's observations relate to appearances only.

Finally, Haller asks, how is circulation established on Wolff's system? Why do the small veins, formed in the outer regions of the area vasculosa, come together in larger and larger branches and finally reach the heart? And how are the arteries also formed in a system parallel and complementary to the veins? "The more I reflect on the developing embryo," Haller remarks, "the more completely I am persuaded that the related parts have been present at the same time; the arteries together with the veins, the organs, the accompanying nerves, and the bones. Chance would never have brought together an artery originating from the heart, a vein continuous with the artery, and a nerve from a completely different origin" (p. 278). The vessels are not formed gradually by an essential force, nor are any of the other organs. "That an animal is built from unformed matter by a single propelling force," Haller concludes, "appears to me to be the same as expecting a river to arise from the Lake of Geneva whose branches have the figure of an eagle" (p. 279).

In an Addenda section to the *Elementa physiologiae*, Haller describes the appearance of the area vasculosa at different hours, remarking "I see all of this [in a manner] similar to what the illustrious Wolff has seen; we differ in our conclusions" (1757–66, 8[1766], pt. 2:218). Reiterating the two experiments he had reported to Bonnet, Haller concludes simply, "Therefore that at all times the vessels have had membranes I believe I am able to take for granted" (p. 219). Haller also added these two experiments to a revised Latin version of his *Sur la formation du coeur dans le poulet*, published in volume 2 (1767) of his *Opera minora*. This version was considerably increased in size, as Haller included observations on chick eggs made during the summers of 1763, 1764, and 1765.

Wolff turned his attention to the area vasculosa again in his work on the development of the intestines (1766–67, 1768), where he presents a significantly more detailed description of the development of the vitelline vessels than he had in his earlier works (see Figure 10, *bottom*). Wolff here introduced the term *area vasculosa*, as he did several others, including *vena terminalis* for the circular boundary blood vessel, altered after Wolff to *sinus terminalis*, the term used today. Wolff continued in this work to argue that the observed development of the area vasculosa is inconsistent with Haller's preformationist description and that it supports instead an epigenetic interpretation.

Years later, in a publication written after Haller's death, Wolff recalled his controversy with Haller over the area vasculosa. "I did not want at that time," he writes, "to dispute against this great and admired man any longer. But it is clear that he, not I, has concluded incorrectly from the appearances that are the same overall for him and for me" (1789:14 n.). Wolff describes the two experiments that Haller had performed to disprove his theory on vessel formation. Yet, Wolff counters, these experiments prove nothing. If Haller was experimenting on the area vasculosa at a time when it possessed colored blood vessels, then the experiments would be expected to turn out as they did. By this stage in development, Wolff claims, vessels do indeed have thickened matter around them, because the heart has begun beating and the impulsion of blood causes an increase in density of the material around the vessel channels. Consequently, the vessels would behave as if they had true membranes. Thus, Haller's experiments really proved nothing against his own theory, Wolff claims, because they were performed too late in the developmental process.

The formation of the heart

Intimately tied to the debate between Haller and Wolff over the area vasculosa was their controversy over the formation or preexistence of the heart. For Haller, the heart was the prime mover of development. At conception, according to Haller, the tiny invisible heart, with its inherent irritability, is stimulated; its subsequent beating produces the gradual unfolding and development of the preexisting embryo. Thus, it was absolutely necessary for Haller to maintain that the heart exists at all stages of development. For Wolff, on the other hand, it was equally necessary to demonstrate that there is *no* heart during the early hours of development, so that conclusive evidence could be exhibited for the operation of the *vis essentialis*. Because the embryo can be observed to begin to develop before a heart can be seen to exist, there must be a force, Wolff concludes, that is responsible for this early movement of fluids. It is not surprising, then, that the issue of the heart's formation became a focal point of the Haller–Wolff debate.

From the *punctum sanguineum* ("point of blood") of Aristotle to Harvey's *punctum saliens* ("leaping point"), the appearance of the heart in early stages of development was a universally noted event among observers of the embryo. Indeed as Adelmann has remarked, "The beating heart, making its appearance while the other parts are still largely so inconspicuous as to escape the untrained observer, is perhaps the most striking, indeed, the most dramatic feature of the early embryo" (1966, 3:1300). Most early embryologists consequently believed that the heart is the first organ formed, and many concluded that it was from the heart that other structures then develop (see Adelmann 1966, 3:1300–1389). As Harvey proclaimed, "In a word I say, – from the cicatricula [blastoderm]... proceeds the entire process of generation; from the heart the whole chick, and from the umbilical vessels the whole of the membranes... that surround it. We therefore conclude that the parts of the embryo are severally subordinate, and that life is first derived from the heart" (1651, 1847 trans.:397). Harvey believed that the heart does not beat before it is filled with red blood – indeed, he argued that the blood is made even before the heart. Others, Malpighi among them, thought that they could observe a heartbeat before the heart is red.

It was again Malpighi who produced the first detailed observations on the structure of the early heart. Although they were not free from error, Malpighi's descriptions of the morphological changes in the embryonic heart, as it develops from a U-shaped tube to a four chambered structure, parallel essentially the gross anatomical descriptions given today. Malpighi believed that the heart, like other structures, preexists before incubation. Yet he also showed that it undergoes complex structural changes before it reaches its final form.

We know now that the heart does not really begin to form until the end of the first day of incubation. By 29 hours, its structure is that of a tube, which, because it is growing faster than the rest of the embryo, soon bends into a U-shape. Then at about 40 hours of incubation, the heart begins to twist into a loop, at the same time that the embryo is twisting onto its side (as it undergoes torsion). After the heart attains its loop structure, the chambers begin to partition off from one another, resulting eventually in a four-chambered structure.

As the structure of the heart develops so do its functional abilities. The first contractions of the heart occur at about 29 hours of incubation, but these are very slow and weak, not capable of setting the blood in motion. As these tentative beats gradually become stronger, the heart's structure develops. Contemporaneously, the vitelline circulatory arc begins to reach completion. Finally, at about 40 hours of incubation, the last link in the vitelline circulation is completed, the heart begins to beat firmly, and circulation commences (see Patten 1971).

Haller informs us in his *Sur la formation du coeur dans le poulet* that it was a logical difficulty that he saw in Malpighi's account of the development of the heart that prompted him to begin his own observations on incubated chicken eggs. Haller could not see how the pulmonary circuit could possibly develop in the chick embryo given Malpighi's description of the heart. Malpighi had believed that, when the heart is a curved tube, one auricle and two ventricles exist in serial order, the ventricles being connected by a short passageway (see Figure 11). Haller could not envision how the pulmonary circulatory arc could insert itself, as it were, between the two ventricles originally connected directly together. "If Malpighi has observed correctly," Haller explains, "by what mechanism could this artery and this vein place themselves between the two ventricles and produce communication, when [according to Malpighi] they have not existed during the first days of the fetus, when the blood of the right ventricle passes into the left ventricle by a canal which admits no vestige of the vessels of the lung? It is to solve this problem that I have commenced my series of observations" (1758a, 1:9).

Haller's solution, worked out as we saw in Chapter 2 during the observations he made on incubated chicken eggs in 1755, 1756, and 1757, was twofold. First, he came to believe that the pulmonary vessels are simply invisible during the first days of development. Second, Haller became convinced that Malpighi had been wrong in identifying two ventricles in these early stages. As Haller reported to Bonnet, "Nothing is simpler than the metamorphosis of the heart; the right ventricle is almost not apparent in the beginning; it comes out in the manner of a bump on the left ventricle, the only original ventricle. Malpighi

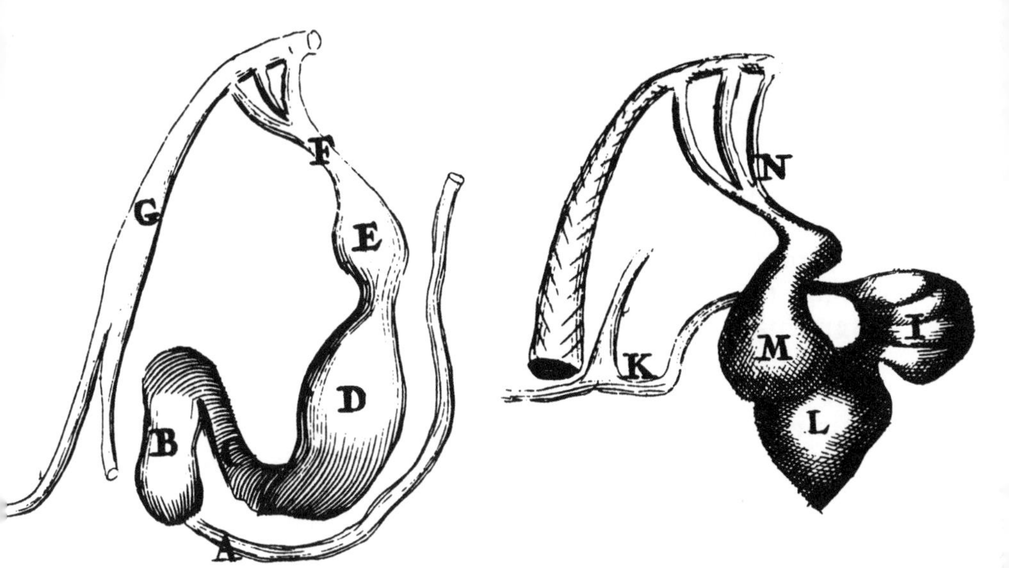

Figure 11. Malpighi's illustrations of the developing heart. *Left*, 40 hours; *right*, 4 days. *B/I*, auricle(s); *D/L*, right ventricle; *E/M*, left ventricle. (From *Opera omnia*, 1686)

has constantly called the left ventricle that which is only the bulb of the aorta, and the right ventricle that which is surely the left" (Bonnet MSS, 8 January 1757). Haller found, through his observations, that initially one can see only one ventricle in the embryonic heart; Malpighi's other ventricle Haller called the bulb of the aorta (a name still used today). The two ventricles never exist connected in a series, as Malpighi thought. Instead, Haller believed the left ventricle is the first to be visible with the right ventricle developing as a "bump" on the left. The right ventricle always exists, in parallel with the left, but it remains invisible in early stages of development. Just so, Haller thought, Malpighi's single auricle really contains two auricles. Finally, Haller came to believe that the pulmonary circuit, initially invisible, develops from branches of the aorta (the aortic arches) and therefore never has to insert itself between the two ventricles.

Haller's solution of this problem was one of the major observational aspects of his conversion from epigenesis to preformation (Chapter 2). As he reported in *Sur la formation du coeur dans le poulet*,

I had myself believed that I had found in the heart of the chicken a proof for epigenesis, and I persuaded myself that a recurved tube could become a muscle with four chambers only by the addition of several new parts: but experience itself has shown me that the changes in this principal organ are only superficial and that they are born from its primordial structure by successive degrees, which is a proof for evolution [preformation] instead of being opposed to it. [1758a, 2:172–73]

Haller now argued that the heart only *appears* to be devoid of its four chambers in initial stages of development. They are there from the very beginning, just as the right ventricle exists while only the left is visible. They attain their visible state as the gradual unfolding process of development takes place.

Haller believed that the heart appears to begin beating at about 48 hours of incubation. "Why did movement begin at this time?" Haller asks, since one observes development in the chick before this time. The answer lies, once again, in the heart's transparency; for "because it was transparent and did not strike the eye in one place, nor in another, nor in its passage from the first to the second," its earlier beating was not visible. "One does not see the wind," Haller declares (1758a, 2:178). The heart exists from the first moments of development, tiny and transparent, but nevertheless capable of functioning to produce the embryo's development.

Wolff's discussions of the heart in the *Theoria generationis* revolve primarily around when the heart becomes visible. In an embryo of 28 hours of incubation, he reports, no heart can be seen (1759:72, § 166). By 36 hours, the heart is visible in the form of a white ring, not connected to vessels and not beating (pp. 72–73, § 167; see Figure 8). Yet "that the embryo is nourished at this time by the substance of the egg is demonstrated by the increase in its volume. . . . it follows that the nourishing particles pass from the egg into the embryo, and that there exists a force, through which this is accomplished, which is not the contraction of the heart and the arteries" (p. 73, § 168). Upon the existence of this force, and the absence of the heart, rests Wolff's whole explanation of development.

In the *Theorie von der Generation*, Wolff repeats that he has seen the heart at 36 hours and that it was not beating. "I am certain," he writes, "that the heart has never moved in these embryos, just as I am certain that there is a time, namely before 24 hours, when there is as yet no heart there at all" (1764:265). Wolff also reports new observations that he made on an egg incubated 29 hours. To his surprise, he could see a small heart. Not only that, but he could observe that it made weak and irregular contractions, which, however, pumped no fluids. Movement, Wolff concludes, "does not arise all at once, as I once imagined, but by degrees; it gradually becomes stronger and at last finally passes over from a slow spastic movement, as I have described it, into an instantaneous, convulsive movement, as we still see in adults" (p. 268). Thus, we cannot only observe the heart forming, we can also view the gradual development of the heartbeat. "Should all this still not be sufficient," Wolff concludes, "to prove my theory of the essential force, of the gradual formation of the parts and the gradual attainment of the characteristics that . . . they must have in adults? I shall leave the whole matter to the judgment of Herr von Haller" (p. 270).

As we have seen, Haller reported to Bonnet that, because of Wolff's book, he felt that it was necessary to repeat his observations on incubated eggs to satisfy himself "on the preexistence of the heart, which this physician claims to have seen born" (Bonnet MSS, 11 April 1765). Yet Haller added little to his discussions of the heart in the *Elementa physiologiae*. He repeats again his belief in the heart's preexistence, and even reports without comment Wolff's observations on the weak beats of the 29-hour heart. Haller simply concludes, "[Wolff] repeats in several places that the heart has definitely not existed before 24 hours. . . . I doubt strongly the truth of this phenomenon. This illustrious man has seen the heart at 29 hours, and I [have seen it] already formed and perfect at 48 hours. If it was not visible earlier, it is because its smallness, its transparency, its whiteness, and its resemblance to a mucous cellular substance concealed it" (1757–66, 8[1766]:116–17).

An issue closely related to this controversy between Haller and Wolff was Wolff's explanation of why animals have hearts and plants do not. Briefly, Wolff argued that the substance out of which plants are formed solidifies much more quickly and to a more rigid degree than does animal substance. Therefore, during development, only parallel vessels have time to form as fluids move through plant substance. But in animals, Wolff claims, branching vessels can form because solidification of structures proceeds much more slowly. Branching vessels, though, necessarily entail one common branch, and this, says Wolff, is the heart (1759:92–95, §§ 215–16; 1764:8–10).

Haller comments, in his reviews of both the *Theoria generationis* and the *Theorie von der Generation*, that he simply does not understand Wolff's argument. Yet Wolff thought a great deal of his explanation for the difference between plants and animals, as will be made evident in the discussion of Wolff's philosophy of science in Chapter 4. That Wolff's argument was lost on Haller is symptomatic of a wide gulf separating the two in their attitudes toward scientific explanation.

Haller's membrane-continuity proof

Haller's membrane-continuity proof of preformation was perhaps his most novel contribution to eighteenth-century embryological theory. Through observations on developing embryos and on eggs before fertilization, Haller thought he could demonstrate that the fetus must exist in the mother before conception and thus that the theory of preformation must be correct. Announced to Bonnet in September 1757, the membrane-continuity proof was first published in Haller's *Sur la formation du coeur dans le poulet.* This is what he says about it there:

> It seems to me almost demonstrable that the embryo is present in the egg and that the mother contains in her ovary everything that is essential to the fetus. Here are the proofs. The yolk is the continuation of the intestines of the fetus; the internal membrane of the yolk is a continuation of the internal membrane of the small intestine; it is continuous with the internal membrane of the stomach, pharynx, and mouth, and with the skin and epidermis: the external membrane of the yolk is the expanded external membrane of the intestines; it is continuous with its mesentery and peritoneum. The envelope that covers the yolk during the last days of incubation is the skin of the fetus. [1758a, 2:186–87]

The yolk sac, Haller claims, is really just a "large hernia" of the intestines. One can conclude, then, that "If the yolk is continuous with the skin and with the intestine of the fetus, it must have existed with it: it is truly part of the fetus. The yolk existed in the abdomen of its mother independently of the approaches of the male; the fetus must have likewise existed there, although invisible and enclosed in an amnion, always apparently upon the yolk, but invisible because of its smallness and its transparency" (p. 188).

Haller's argument has been illustrated by F. J. Cole (1930), reproduced here in Figure 12. Haller claims that one can observe that the internal and external membranes of the embryo's intestines are continuous with the internal and external membranes of the yolk sac. One can also observe, according to Haller, these same membranes covering the yolk sac in the

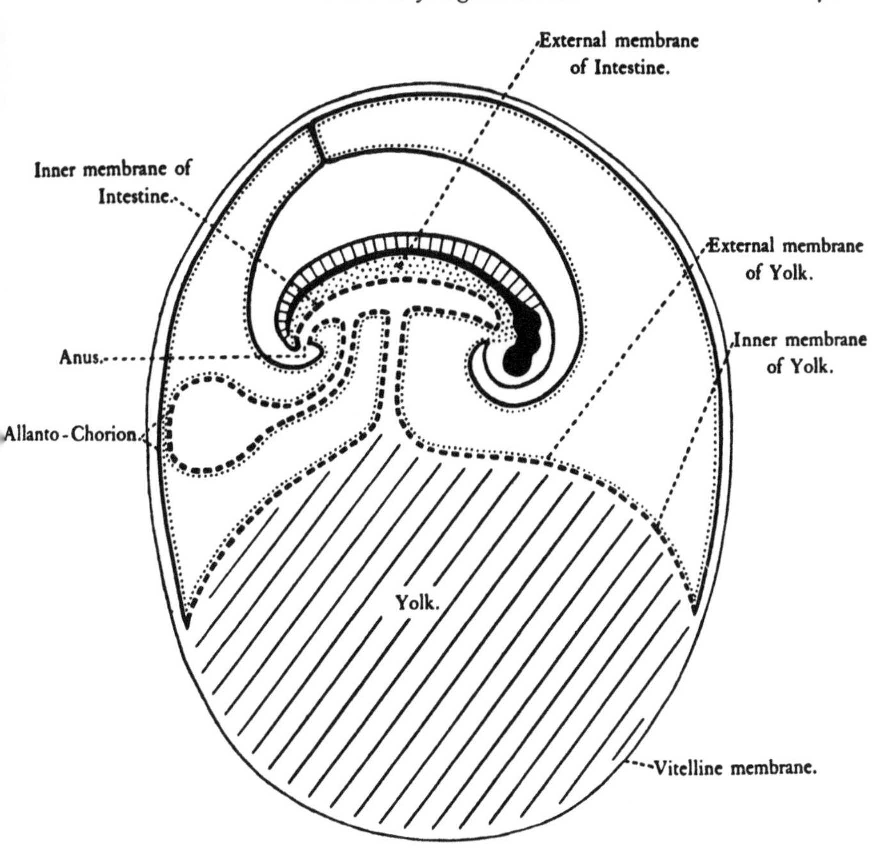

Figure 12. Haller's membrane-continuity proof, as illustrated by F. J. Cole. At this stage of development (about 4 days), the inner and outer yolk-sac membranes clearly extend into the embryo's intestinal membranes. Cole does not note the contemporaneous shrinking away of the vitelline membrane (see Figure 13). (From *Early Theories of Sexual Generation*, 1930; courtesy of Oxford University Press)

unfertilized egg still in the mother. Therefore, he concludes, the embryo must be there also, connected to the yolk sac through its intestines, but too small and transparent to be visible. It is thus demonstrated that the embryo preexists in the mother.

Haller's proof was actually based on faulty observations. Before fertilization, the egg of the chicken is indeed covered by a membrane, the vitelline membrane, which surrounds the yolk substance, the cytoplasm, and the nucleus. After fertilization, the ovum acquires the albumen (egg white) and the shell. And as the embryo develops, the yolk-sac membranes and the embryo's intestinal membranes can indeed be seen to be continuous. But what Haller missed was that the yolk-sac membranes of the developing embryo are different from the vitelline membrane that covers the unfertilized egg, and that the vitelline membrane actually shrinks away as the yolk-sac membranes spread further and further over the yolk (see Figure 13). Thus, although the embryo is connected through its intestinal membranes with the yolk-sac membranes, it is not continuous with the original vitelline membrane.

Haller's error is certainly understandable when one considers what was known among eighteenth-century embryologists about the membranes covering the fetus and the yolk sac. There was considerable confusion over the relationship between the blastoderm, the yolk-sac membranes, and the vitelline membrane. No one before Haller had described the situation any better than he. It was really not until Wolff's observations on the formation of the intestines, published in the late 1760s, that some clarification of the relationship between the embryo and the yolk sac was established.

The origins of Haller's "proof" of preformation are not entirely clear. We know that he devised the membrane-continuity proof during the observations on incubated chicken eggs he made in 1757, although there are no hints of it before his letter to Bonnet on 1 September. Additionally, Haller refers in *Sur la formation du coeur dans le poulet* and in the *Elementa physiologiae* to the work of Antoine Maître-Jan when discussing the membranes of the yolk sac. Maître-Jan had, in his *Observations sur la formation du poulet* (1722), reported extensive observations on the yolk sac and its membranes, and on the yolk stalk (the short canal that connects the embryo's intestinal region with the yolk sac). More significantly, Maître-Jan had suggested an argument similar to Haller's. As Maître-Jan explained, "Since the little white body is the chick in abridged form [*en abrégé*], the parts of which unfold gradually and are fashioned as the egg is incubated, . . . and since the exterior membrane of the yolk is an extension of the peritoneum, and the interior one, covering the humor of the yolk, is an extension of the common membrane of the intestines, there is every reason to believe that this little white body and these membranes are formed at the same time in the ovary" (1722, trans. Adelmann 1966, 5:2167). Maître-Jan believed that the chick develops from a "little white body"[10] that can be observed in unincubated eggs floating on top of the yolk. After the embryo begins to develop, Maître-Jan claimed, one can see that the yolk membranes are really just extensions of membranes of the chick, for the inner yolk membranes are continuous with the intestines and the exterior membranes with the peritoneum (the abdominal lining). On the basis of these later observations, one can conclude, Maître-Jan proposed, that the little white body (the rudimentary embryo) and these yolk membranes were all formed at the same time in the ovary of the female. "And in fact," he maintained, "when the smallest eggs in an ovary are examined, they are found to be white and covered by very delicate membranes" (p. 2167).

Although Maître-Jan's argument is certainly similar to Haller's, Maître-Jan did not carry it to the extreme that Haller did with regard to proving preformation. In his *Observations sur la formation du poulet*, Maître-Jan was most concerned to refute the animalculist view of preformation and to offer instead arguments in favor of ovism. He believed that the little white body could be shown to exist in the unincubated egg and, by inference, in the ovary as well. Yet Maître-Jan did not use his argument to support the preexistence of germs through *emboîtement*, claiming only that the little white body and the yolk membranes are formed at the same time in the ovary. Nor did he ever clearly state that the little white body is a complete miniature of the embryo.

There is no doubt that Haller was fully aware of Maître-Jan's work. Referring to Maître-Jan's book more than once in his *Sur la formation du coeur dans le poulet*, Haller notes with regard to the structure of the yolk that "Maître-Jan is almost the only author who has made any progress in discovering this" (1758a, 2:138). It is quite possible that, in the summer of 1757, when Haller began to investigate the yolk, he turned to Maître-Jan's book for guidance and that, in consequence, he repeated Maître-Jan's observations on unincubated and unfertilized eggs. Although Haller made significant advances in his own observations on the yolk and its relationship to the embryo over those of Maître-Jan, he was sufficiently impressed with Maître-Jan's continuity argument to see its use as a major proof of preformation. Never does Haller hide his debt to Maître-Jan, who is always duly footnoted when Haller discusses his membrane-continuity proof; yet neither does Haller ever contradict those who, like Bonnet, were to ascribe this proof to Haller alone.

Whether Haller was either entirely correct or entirely original with his membrane-continuity proof is less significant than the importance he and others attached to his proof that the embryo is preformed in the mother. As Bonnet declared in his *Considérations sur les corps organisés*, "I have said . . . that one day we would extract from Nature its secret. One of its most cherished favorites, M. le Baron de Haller, has recently interrogated it as it demanded to be, and he has obtained from it responses that extend the boundaries of our knowledge. It is the interior of an egg of the chicken that has rendered to him its oracles" (1762, 1:124). Bonnet enthusiastically endorsed Haller's membrane-continuity proof as a truly great discovery: "I had not hoped," he confessed to Haller, "that the secret of generation would be discovered so soon" (Haller MSS, 30 October 1758).

Yet Haller's proof also had its critics, notably Wolff. In his *Theorie von der Generation* Wolff severely criticized Haller's argument both on logical and on observational grounds. To begin with, Wolff challenged Haller on the relationship between continuity and simultaneous creation. As he queries, "if . . . to put it briefly, the yolk of an egg continues into the embryo – can one, I say, conclude from this that the two elements, the egg and the embryo, therefore necessarily must have begun to exist simultaneously at any given time, either before or after copulation or at the creation?" (1764:105). Can one infer the constant coexistence of two things from their observed continuity? Wolff's answer is no. In his own account of development he had shown that parts of the embryo are secreted from other parts, for example, the wings of the chick from the spinal column. Futhermore, the two structures thereafter remain connected together. Yet the wings and spinal column, Wolff asserts, were clearly not formed at the same time but rather one after, and out of, the other. As Wolff challenges, "I do not see at all the connection between these propositions: A part of a thing is directly continuous with the other part of the same thing; therefore the one part of it could never have existed without the other part, but the whole thing must have been produced instantaneously" (pp. 105–6). Consider, Wolff says, a wall built from bricks and covered with a common coating. After the wall is completed all its parts are continuous with one another. But were they necessarily produced all together at the same time? Clearly not, Wolff claims, and thus Haller's entire proof rests on a logical flaw.

In an appendix to his *Theorie von der Generation*, Wolff returns to Haller's membrane-continuity proof, criticizing it now on the basis of new observations he had made on chicken eggs. Haller's proof, he correctly states, rest ultimately on the assumption that the membrane that covers the yolk sac and is continuous with the embryo's intestines is the *same* membrane as the one that covered the yolk before fertilization. "But," Wolff asks, "is it certain that the membrane of the yolk, with which the embryo is continuous, is already present before incubation? It is precisely in this certainty that the entire proof drawn from the observation should consist" (1764:274). Yet Wolff claims that his observations show that this is not the case – that the two membranes are indeed *not* the same. "I am ashamed," he admits regarding his earlier discussion, "of the ridiculous dispute that I carried on in such detail, so clearly and so plainly, against the conclusion based on continuation; for such a continuation has never in the world existed in eggs; and this time I am like the scholars of old who disputed so ardently about the way the golden tooth arose" (p. 275).

Before presenting Wolff's new observations, let me review briefly the sequence of formation of the various embryonic membranes in modern terms, so that the full extent of the Haller–Wolff exchange on this subject can be appreciated. Before fertilization, the yolk, cytoplasm, and nucleus are enclosed in a simple membrane, the vitelline membrane. Following fertilization, the albumen (egg white), the shell membrane, and the shell are added, and the egg is laid by the hen. As the embryo begins to develop (one can picture it as sitting on top of the yolk, as it were), a membrane called the serosa (or, alternatively, the chorion) begins to spread out from the embryo over the whole embryo–yolk complex, while at the same time the yolk-sac membrane (together with the area vasculosa) spreads out from the bottom of the embryo over the yolk (see Figure 13). As this happens, the original vitelline membrane shrinks away before the advancing serosa and yolk-sac membranes. A third major structure then begins to develop, the allantois, a sac that serves to collect wastes from the embryo. The allantois continues to expand until it surrounds the entire embryo and yolk, and its outer membrane fuses with the serosa (allowing oxygenation of the embryo's blood). The two joined membranes are now called the chorioallantois. While all of this has been happening, the albumen has decreased in bulk so that it now occupies a very small portion of the egg. Consequently, the chorioallantois (the fused serosa and allantoic membrane) is now right next to the shell membrane, and the embryo, yolk sac, and allantois, which it contains, occupy almost the entire egg. All of this takes place by about the 14th day of incubation (Hamilton 1952; Patten 1971).

Wolff reports from his observations that when one opens an egg incubated for 12 to 16 days, carefully removing the shell, one first encounters an opaque membrane surrounding the

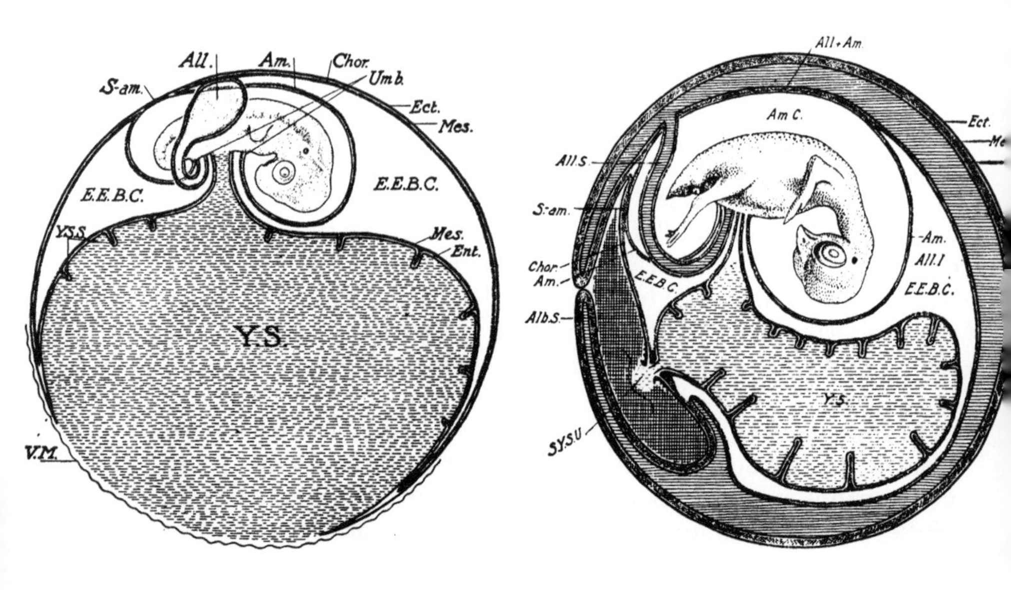

Figure 13. The development of the embryonic membranes. *Left*, 4th day of incubation. Note the serosa (*Chor.*) and yolk-sac membranes advancing over the yolk (*Y.S.*) as the vitelline membrane (*V.M.*) shrinks away. *Right*, 12th day of incubation. Note the expansion of the allantois (*All.*) to surround the embryo in the amnion (*Am.*), the yolk sac (*Y.S.*), and the albumen (*Alb.S.*). The outer chorioallantoic membrane is here designated by its three layers, ectoderm (*Ect.*), mesoderm (*Mes.*), and entoderm (*Ent.*). (From Hamilton, *Lillie's Development of the Chick*, 1952; courtesy of Holt, Rinehart and Winston, Inc.)

entire contents of the shell (the shell membrane). If this membrane is peeled away, a second, transparent one is found underneath, which Wolff identifies as Malpighi's chorion (most likely the chorioallantois). This transparent membrane, Wolff describes, encloses the embryo, the yolk sac, and the albumen. It is this membrane, Wolff claims, that should be identified with the membrane that covers the unfertilized egg. Here is why: if one removes this transparent, common membrane from the incubated egg, one finds that the embryo (enclosed in the amnion) and the yolk sac are connected together at the embryo's intestines via a short canal (the yolk stalk). Furthermore, one can see that the yolk-sac membranes and those of the embryo's intestines are indeed continuous, as Haller had maintained. Yet it is impossible that these yolk-sac membranes could be the original membrane that covered the unincubated egg; for as development proceeds, the embryo forms *within* this original membrane. Thus, the yolk-sac membranes must form at a later point in development, not before fertilization.

Wolff was in many respects correct in his criticism of Haller's proof, and his observations provided a step forward in the clarification of the relationships among the various membranes surrounding the developing chick. The yolk-sac membranes are indeed not the same as the original membrane (the vitelline membrane) that covers the unfertilized egg. Wolff was correct in pointing out that the embryo begins to develop within this original membrane and can therefore not become continuous with it via its intestines. What Wolff did not realize was that this original, vitelline membrane is also not the same as the one he found enclosing the embryo, yolk sac, and albumen later in development (the chorioallantois). As was mentioned earlier, the vitelline membrane actually shrinks away during development as the serosa and yolk-sac membranes spread further out over the surface of the yolk. Yet even with this error, Wolff's observations dealt a blow to Haller's membrane-continuity proof. But Haller remained undaunted. After noting Wolff's arguments in a letter to Bonnet, Haller simply remarked, "he admits all my facts, upon which consequently one can depend" (Bonnet MSS, 18 February 1765). And in his triumphant letter of 25 August 1765, where he described to Bonnet his two experiments on blood vessels in the area vasculosa, he noted,

> Concerning the question, do the parts of the egg exist in the mother; are they joined there with the embryo? The membrane of the yolk comes certainly from the mother; it is found as well in the unfecundated egg; and it is thus not a production of the sperm [i.e., of fertilization]. Since it forms very surely the passage to the yolk, and as this passage holds to the skin and the epidermis of the fetus, this invisible embryo is continuous with the yolk. It seems to me that this is all there is to say. [Bonnet MSS]

By the time Haller published the revised Latin version of his *Sur la formation du coeur dans le poulet* in 1767 (in his *Opera minora*), he had reinvestigated the embryonic membranes in the chick, reporting numerous new observations on incubated eggs. Haller was now able to clarify the relationship between the allantois and the area vasculosa. He distinguished the allantoic blood vessels as an independent system from those of the area vasculosa, and he offered a fairly complete description of the entire development of the allantois. Haller now reserved the term "umbilical membrane" for the allantois, calling the area vasculosa the "umbilical area."

Although Haller footnoted Wolff's observations on the embryonic membranes in the revised edition of his chick memoirs, he did not respond to them. But in the *Elementa physiologiae*, Haller used his own new observations to challenge Wolff's conclusions. Having observed the expansion of the umbilical membrane (the allantois) to fill the egg shell and completely surround the embryo and yolk, Haller felt that he was able to undercut Wolff's argument. Although Haller's comments are brief, they suggest that he thought that the membrane Wolff identified as containing the embryo, yolk sac, and albumen at 12 to 16 days of incubation was actually the umbilical membrane (allantois), newly expanded to fill the shell. Consequently, Wolff's having found this membrane at such a late stage in development could not diminish Haller's own argument, because Wolff was in error in identifying this newly expanded membrane with the original membrane covering the ovum before fertilization. Thus, the yolk-sac membranes that are continuous with the embryo's intestines could still be identical with this original membrane, as Haller had initially claimed in his membrane-continuity proof. "This is why," Haller concludes, "in the first place the fetus is entirely part of the yolk, which is not disavowed by this illustrious man; second, even in the mother the fetus was part of the egg, since the membrane of the yolk was also in the mother and was connected by an indivisible bond with the intestines and the whole chick" (1757–66, 8[1766]:95).

A major factor in the controversy between Haller and Wolff over Haller's membrane-continuity proof was certainly that neither had all the embryonic membranes and structures correctly identified. Haller was indeed in error in identifying the yolk-sac membranes of the incubated chick with the vitelline membrane of the unfertilized ovum. Wolff was correct in his criticism of this aspect of Haller's argument but for the wrong reasons. He did not see the actual vitelline membrane shrinking away as the serosa and yolk-sac membranes advance. Rather Wolff thought the vitelline membrane remained and that all the embryonic structures (amnion, yolk, albumen) develop within it. Here Haller made the more correct observation, for he finally observed the allantois expanding around these other structures. The transparent membrane that Wolff found under the shell membrane in the eggs of 12 to 16 days' incubation (the chorioallantois) could thus be explained by Haller as a later formation. Wolff's observations did not completely undercut Haller's membrane-continuity proof; for Haller was able to account for the appearance of this new encompassing membrane as a product of the expanding allantois.

Had Wolff also been able to show that the original vitelline membrane covering the unfertilized ovum shrinks away during development, he would have been able to prove conclusively that Haller's membrane-continuity argument was wrong. Yet Wolff's original objection, that the continuity of two things does not entail their simultaneous creation, was never fully answered by Haller, even though he produced arguments to show that the embryo is not grafted on to the yolk during development. Wolff's logical argument points up the real flaw in Haller's membrane-continuity proof, for Wolff claimed that Haller had not really proved anything. He may have demonstrated, if the embryo really does preexist in the mother, where it is most likely located and how its later structure would be consistent with its earlier, transparent one; but all of this assumes that the embryo is indeed preformed, in an invisible state, in the mother. Thus, even though the observational dual between Haller and Wolff on this point was something of a draw, Wolff's logical arguments would seem to have tipped the balance in his favor.

THE FORMATION OF THE INTESTINES

After publishing the *Theorie von der Generation* in 1764, Wolff continued to make observations on incubated chick eggs. The result of these was his most important piece of observational embryology, "De formatione intestinorum," which was published after Wolff moved to St. Petersburg, in the Academy of Sciences' journal for the years 1766–67 and 1768. Although Wolff did not emphasize it, one can surmise a definite connection between his previous work and his observations on the intestines; for it is a natural step from his controversy with Haller over the yolk-sac and intestinal membranes to consider the formation of the intestines themselves.

Wolff admits in the opening section of "De formatione intestinorum" that in his previous work on generation he had been able to explain development in only a very general way. No one in fact, he claims, had ever fully explained the formation of any single part of the embryo. "I am therefore offering this illustrious Academy," he proclaims, "the first specimen of this kind of investigation. I believe that I have detected the very origins of the intestinal canal so clearly that I am able to explain how it arises out of its first beginnings and is gradually perfected and brought to full growth" (1766–67:404; trans. Adelmann 1966, 4:1658).

In his earlier works of 1759 and 1764, Wolff had maintained simply that the internal organs, intestines included, were formed from material secreted from the spinal column, which then solidifies and acquires vessels and vesicles in the process. Yet by 1767, he was able to show that the intestinal canal develops from an originally flat membrane that gradually acquires the form of a tube. Wolff was thus the first to describe the formation of embryonic structures by the folding and fusion of tissue, a process that occurs repeatedly in embryological development.

In modern terms, intestinal development proceeds as follows: Initially, the embryo begins as an elongated, flat region of tissue on top of the yolk. During the second day of development, tissue begins to fold under at the head region, followed by a similar folding under in the tail area beginning on the third day. As this process continues at each end, the embryo, in longitudinal cross-section, begins to look something like a mushroom on a stalk (see Figure 12). In the head region, the tissue that has folded under forms a pocket called the foregut (in which the pharynx, esophagus, and stomach later develop). The tail fold creates the hindgut (in which the intestines and other structures develop). And between them, where the embryo still opens directly onto the yolk, is the midgut, connected to the yolk membranes via the yolk stalk. After these regions are formed, the organs themselves begin to develop during the later days of incubation.

Wolff observed his specimens from the bottom, that is, from the yolk up. He described accurately the formation of the foregut and hindgut by the swelling of tissue in the head and tail regions, and the lateral folding of tissue that constricts the midgut region and the yolk stalk (see Figure 14). Wolff observed this process through the 4th day of incubation, noting the gradual formation of the stomach, the intestines, and the rectum, and the further constriction of the yolk stalk as development proceeds.[11]

Wolff frequently stressed the significance of his observations for the preformation–epigenesis controversy. "When the formation of the intestine in this manner has been duly weighed," Wolff proclaims, "almost no doubt can remain, I believe, of the truth of epigenesis. For if the intestine is at first a simple membrane which then folds so that it becomes double . . . I am certain that this intestine has obviously been formed, that it has not lain hidden for a long time complete and whole and only now

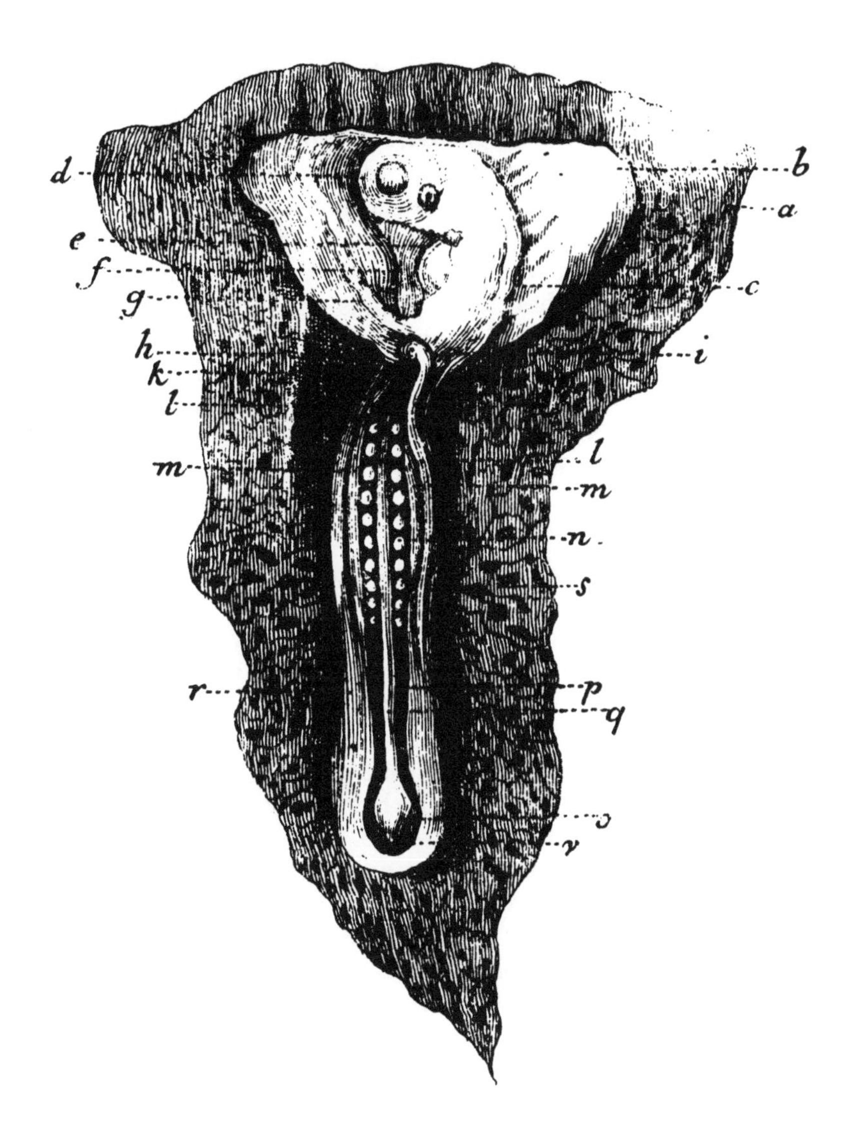

Figure 14. Wolff's illustration of the formation of the intestines at 54 hours (viewed from below): *b*, head sheath (head fold); *e*, heart; *h*, origin of the head sheath; *k*, fovea cardiaca (anterior intestinal portal); *m*, interior intestinal borders; *n*, rudiments of the vertebrae; *q*, spinal cord; *v*, first outline of the tail fold. (From "De formatione intestinorum," 1766–67)

come to view" (1766–67:460–61; trans. Adelmann 1966, 4: 1667). If the stomach, for instance, were preexistent, only gradually becoming visible as it unfolded, we would have to be able to see it, Wolff argues, as a whole stomach from its first moments of visibility. If it were preformed, as soon as it could be seen, "however small it might be," it "would have to have the true form and appearance of the adult stomach; but it would never be half a stomach, would never be open, and would never be joined with parts not belonging to it" (p. 455; trans. Adelmann, p. 1664). Yet one can observe the stomach in these half-formed stages; therefore, organs do not exist in a tiny, complete form, expanding during development to their adult size. "This, I say," Wolff asserts, "does not seem to be Nature's design, but she has intended rather that the formation of organic bodies should in general be left solely to the natural forces inherent in animal and vegetable matter" (p. 455; trans. Adelmann, pp. 1663–64). Nature dictates epigenesis, not preformation.

Haller reviewed Wolff's "De formatione intestinorum" in the *Göttingische Anzeigen von gelehrten Sachen* for the years 1770 and 1771.[12] Both reviews contain Haller's customary summary, and it is only at the conclusion of the second review that we find any reaction from Haller. Here he challenges, "But should it not remain possible that a highly transparent intestine has appeared to be simple and flat under the magnifying glass when it was already actually closed? At least the beginning of the thorax is so extremely transparent, that it would be invisible without the acid that brings it out, and one day earlier this beginning can be entirely present indeed, and yet cannot be brought to visibility by any acid" (1771:415). It is not surprising that Wolff's account would be dismissed by Haller in this way; the defense of transparency was used by Haller throughout the controversy to undercut true development of parts. Yet one cannot help feeling a sense of regret that Wolff's remarkable observations should have been relegated to the class of optical illusion by Haller, for they were indeed a milestone in descriptive embryology.

Wolff noted Haller's less than enthusiastic response to his work in a letter to Leonard Euler, written in 1783, six years after Haller's death. The subject was the appearance of a new edition of Bonnet's *Contemplation de la nature* in 1781 (in Bonnet's *Oeuvres*), which contained some additional footnotes on preformation. Wolff refutes each of these in turn, particularly one on the membrane-continuity proof, "to which," he writes, "twenty years ago I responded two things: 1) that it [the continuity of membranes] does not exist as it is said, and as it would have to appear in order to prove evolution; [and] 2) that if it existed, as the late M. de Haller would have it, it would prove nothing." Wolff continues, "The solid refutation of my objections by the late M. de Haller consists of an observation by which he believed he had seen that at a certain time of incubation (it could only have been at the beginning of the third day), the vessels of the umbilical area exist already all formed. M. de Haller (in a letter to M. Bonnet) added a single word of scoffing. He believed, he said, that after this I would be quieted down." This is indeed what Haller had written, in his letter to Bonnet (25 August 1765) announcing his two experiments on blood vessel formation, later published by Bonnet in his *Oeuvres* (1779–83, 5[1779]:278–79). "But," Wolff continues to Euler, "I wrote after this my two dissertations on the formation of the stomach and the intestines. M. de Haller has called them very important, and he has never responded to them. It seems, then, that M. de Haller was quieted down himself" (Castellani 1971:513).

Wolff's "De formatione intestinorum" caused little stir among his contemporaries. In fact, it was not until J. F. Meckel translated and published it in 1812 as *Über die Bildung des Darmkanals im bebrüteten Hünchen* that Wolff's work on the intestines became widely known. Its influence was particularly marked on Pander and von Baer, both of whom recognized the germ layers, a concept toward which Wolff's work was a definite step.[13] Von Baer, in fact, hailed Wolff's work as "the greatest masterpiece we know of in the field of the observational sciences" (1828–37, pt. 2:121 n.).

OBSERVABILITY AND EXISTENCE

A constant subtheme in the exchanges between Haller and Wolff on the area vasculosa, the heart, the membrane-continuity proof, and the formation of the intestines was the issue of the relationship between observability and existence. Can one postulate that an embryonic structure does or does not exist if it cannot be seen? Clearly this is a central issue for embryology, particularly in the seventeenth and eighteenth centuries when good compound microscopes were not yet available (see Bradbury 1967). Simple microscopes, and some compound, were indeed in use, but the earliest stages of embryological development remained hidden from view. It is pertinent in this regard to note Gasking's argument (1967:45–46) that the invention and use of the microscope in the seventeenth and eighteenth centuries indirectly aided the preformationists' position by demonstrating that minute structures do indeed exist that are invisible to the unaided eye. As a result, the possibility that even smaller structures might exist, not yet made visible by microscopes, became much more plausible (see also Bodemer 1964:23).

Indeed, Malebranche's first enunciation of the *emboîtement* theory in his *Recherche de la vérité* was made within the context of a discussion of the limits of our senses. There Malebranche noted that microscopes had revealed animals much smaller than had ever been imagined; perhaps "there could be smaller and smaller animals to infinity, although our imagination is alarmed at this thought" (1674:81). With lenses one sees the germ of the tulip in the bulb, and one can conclude "the same thing . . . in general of all sorts of trees and plants, although this cannot be seen with the eyes or even with the microscope" (p. 82). The existence of germs encased within one another may be beyond the realm of our senses, Malebranche admitted, "But it is not necessary that the spirit stop with the eyes: for the view of the spirit is much further than the view of the body" (p. 83).

The concept of *emboîtement* could be accepted only on this basis, that dimensions of size far smaller than we can even imagine must be possible. The use of the microscope aided this notion by revealing a world below our senses. Microscopes furthered the preformationist cause also by making visible smaller and smaller parts within embryos. That these parts could all have existed in an invisible state since the Creation became a plausible idea.

A common technique used to aid visibility was to apply vinegar or alcohol to embryonic fluids, causing structures to appear through the coagulation of the viscous fluids.[14] In many cases, artifacts were undoubtedly produced by this method, but organs were indeed made visible slightly earlier than normal through this technique. Haller often spoke of seeing the lung, for instance, a day earlier than usual by applying vinegar or wine. Wolff also used this technique but claimed that it was an aid to observing existing parts that are hidden only because of temporary transparency.

Epigenesists had to maintain, of course, that one can prove that embryonic parts are newly born during development and that they do not exist before this time. Wolff addressed himself to this issue in his dissertation:

In general, it cannot be affirmed absolutely that whatever is not accessible to our senses therefore does not exist. However, applied to these experiments, this principle has in fact more elegance than truth. The constitutive parts, of which all parts of the animal body are composed in their first state, are globules that always yield to [i.e., are discernible by] a mediocre microscope. But who would say that he was unable to see a body because of its smallness, when nevertheless the parts of which it is composed are precisely because of their smallness unable to escape notice? No one has ever uncovered with the help of a stronger lens parts not also detected immediately with a microscope of cheaper quality. Either they are detected through no means, or they appear sufficiently large. Therefore, that parts are hidden because of their infinite smallness, thereafter gradually emerging, is a fable. [1759:72, scholium to § 166]

Wolff believed that all parts in plants and animals, as well as the substance out of which these parts form, are composed of globules (not cells in the modern sense).[15] Because one can always observe these globules, he argued, which are certainly smaller than the organs they make up, one can conclude that if a larger organ is not visible, it does not yet exist.

Haller noted this argument in a letter to Bonnet, written when he was reading Wolff's dissertation for review. Here he remarks, "This Wolff, defender of Epigenesis, maintains that all the parts of the embryo are composed of globules. Is this really true? I have indeed seen globules in the blood, grains in the umbilical area, types of balls that are the germs of the vertebrae, but have you ever seen the cellular tissue, [or] the membranes in the form of globules?" (Bonnet MSS, 3 October 1760). In his review, Haller elevates Wolff's observability argument to the status of the foundation of his system, claiming that "In the production of animals, one must pay attention to a principle, which stands directly at the beginning [of the section on animals], and according to which that which is not there, one does not see. The basis for this is, according to Herr Wolff, that everything in animals arises from little globules; but these are visible, consequently no part can be assumed [to exist] which is invisible and still present" (1760:1228–29). "But," Haller responds, "whoever has practiced much himself with the magnifying glass will have informed himself . . . that indeed strong colors make parts visible and transparency makes them invisible" (p. 1229). Even in the adult animal, parts are sometimes invisible because they are transparent. But this does not allow one to conclude that they are not there.

Wolff's response to Haller's challenge was to argue, first, that his discussion of the relationship between observability and existence was put forward in a scholium. It could not be a fundamental principle or axiom, because a scholium is by definition not part of the main argument. Wolff proceeds to reiterate his globule argument, claiming that "the heart, the vessels . . . , the wings and feet, these parts, which all consist of globules, [and] which are actually only white and indeed somewhat as all bodies when they are small and thin, but in no way are completely transparent as water or crystal, that these parts because of their smallness or transparency could perhaps remain invisible, appears to me, if I should say as I think, to be an egotistic excuse" (1764:91).

Second, Wolff maintains that such a principle as "what one does not see, is not there" is not even valid. If something is not visible, then one cannot conclude anything about it at all, because one has no grounds on the basis of which to draw a conclusion. But one can prove existence or nonexistence from corroborative evidence. And this Wolff believed he had succeeded in doing with regard to the blood vessels in the area vasculosa. "Transparency," Wolff asserts, "is only something one is forced to take refuge in. I adduce only observation" (1764:118). The preformationists, not he, had erred on the observability–existence issue. "In short," Wolff remarks, "the whole story of invisibility is only a chimera. It can cause confusion, but if we build anything upon it, it always deceives us" (p. 127).

Haller's response to Wolff's remarks was simply to repeat his transparency argument in the *Elementa physiologiae*. It is certainly true, he admits, that the embryo appears to form out of an unorganized primitive glue, but it does not follow that this glue does not contain the parts of the embryo. "I have often given consistency to this jelly," Haller claims, "with spirit of wine alone, in order [to show] that what appeared to be pure jelly, was fibers, vessels, and viscera" (1757–66, 8[1766]:116). Structures do exist in transparent states; one can prove that this is so. Consequently, Haller declares:

> No part therefore in the animal body has been made before any other, and all created parts exist at the same time. If some have said that the first origin of the new animal is the spine, the brain, or the back, or the heart; if Galen taught that the liver is the first to be produced; if others regard the abdomen with the head, [or] the spinal cord and the brain, as first, with the condition that the remaining parts are produced from these, I believe that these illustrious men mean nothing more than that the heart and the brain are visible to the eye at a time when the remaining parts lie hidden: and that some parts of the body are developed to such a degree in the first days of the fetus to be perceived. [p. 148]

All is apparent development only; no true epigenesis takes place.

The issue of the relationship between observability and existence was certainly not one that could be settled, with any amount of argument, between Haller and Wolff. Their differing viewpoints on this subject, as on other key points of contention, were too fundamental to their separate positions to be alterable.

THE EMBRYOLOGICAL DEBATE IN RETROSPECT

In considering the Haller–Wolff debate in general, one is struck by how inconclusive it was. None of the three principal issues of controversy was really settled one way or another, and the debate itself seems to have ended principally because Wolff moved to Russia and Haller ceased doing embryological research. Of course, one could argue that the issues that concerned Haller and Wolff could not be conclusively cleared up during their day because of a lack of full observational knowledge of embryological development, due in part at least to poor optical equipment. Yet this does not seem to be the complete answer, for there is no doubt that if, for instance, the formation of blood vessels in the area vasculosa had been fully described in their day, Haller and Wolff would still have disputed about how this description was to be explained. It is on the level of explanation that one must seek the roots of the inconclusive nature of the debate, and even the source of the controversy itself.

In one of his letters to Haller, written in response to receiving the eighth volume of Haller's *Elementa physiologiae*, Wolff commented:

As far as our controversy is concerned, I think this way. The truth is no dearer to me, illustrious one, than to you. Whether organic bodies of nature are evolved from an invisible state to a visible one, or are truly produced, there is no reason why I should choose this one rather than that or why I should prefer that one, not this. And the same is also your view, illustrious one. The truth alone we both pursue; we search for what is true. Why then should I dispute against you? Why should I resist you, when you strive with me toward the same goal? I would rather entrust epigenesis to your care, to defend and perfect, if it were true; but if false, it would be an odious monster to me also. I would admire evolution [preformation], if it were true, and I would worship most submissively the revered Author of nature, the Divine Power which is inexplicable to the human intellect: but if false, you also would quickly reject it, even if I were silent. [Haller 1773–75, 5:292; letter of 6 October 1766]

Wolff was apparently quite sincere in his claim that each would give up his theory if the other proved to be the truth. Yet one cannot imagine this situation actually arising, for the essence of debate is that neither party is in a position to fully recognize the possible truth of the opponent's theory (see Chapter 6).

While the embryological controversy between Haller and Wolff ensued, a deeper philosophical war was waged simultaneously, which provided the key motivation behind the debate itself. What Haller and Wolff were really arguing about is how one *ought* to explain embryological development, and it was because each had a set of criteria for scientific explanation that was markedly different from the other's that their disagreement ensued. It was the underlying philosophies of science of Haller and Wolff that were responsible for their debate. And it was because of this that their controversy remained inconclusive, for both were starting from very different philosophical positions and reaching observational conclusions on the basis of them. Consequently, it was not the observations that were truly under contention, but rather their ties to this nonobservational plane of controversy.

4

The philosophical debate: Newtonianism versus rationalism

Haller and Wolff present many contrasts to each other. At the start of their debate Haller was fifty-one years of age, an established and renowned scientific figure, whereas Wolff was a twenty-five-year-old graduate of medical school. Wolff was educated in Germany and traveled little until his move to St. Petersburg eight years after he received his degree. Haller received his training in Switzerland and, most significantly, in Holland; and he traveled to and corresponded with scholars in all of the principal European countries. But of most importance for our purposes here, Haller and Wolff were heirs to markedly different philosophical traditions – traditions that had major impact on their scientific work. Haller's adherence to Newtonian mechanism and his strong religious beliefs contrasted sharply with Wolff's emphasis on logic and rationalism, and his negative views on mechanical reductionism. These philosophical differences played fundamental roles in the embryological work of Haller and Wolff, and in the debate that ensued between the two. For it was as a confirmed Newtonian that Haller turned to preformation in search of an explanation for embryological development that would be consistent with both his mechanistic and his religious views. Likewise, it was Wolff's rationalist and antireductionist perspective that led him to epigenesis as the more acceptable explanation of generation. The debate that arose between Haller and Wolff can thus be viewed as the tip of the iceberg, as it were; for the embryological positions of the two physicians were in large measure dictated by their more hidden philosophical persuasions.

In order to discuss this contention more fully, I shall present first a synoptic view of the philosophies of science of Haller and Wolff, drawing upon their various publications to illustrate their basic beliefs. Following this, I shall discuss the debate between them once more, summarizing the philosophical issues involved and the differing roles these issues played as the debate progressed.

HALLER'S PHILOSOPHY OF SCIENCE

Haller was not only an anatomist and physiologist but also a practicing physician, a botanist, a university professor and academician, a civil servant and politician, and a bibliographer and reviewer of scientific, political, philosophical, and literary works.[1] In addition to his scientific publications, Haller wrote poetry, composed three novels, and published works on political and theological topics. Hailed as the "last universal scholar" (Toellner 1971), Haller produced a total bibliography that is truly staggering.[2]

Yet, despite this prodigious output, one can identify a number of themes in Haller's thought that recur again and again in his works, acting as guiding principles in both his scientific and nonscientific writings. Among these, the most important for our purposes are Haller's religious beliefs, his empirical and antirationalist leanings, and his adherence to a mechanical view of physiological explanation. Let me discuss each of these in turn.

The world as God's creation

Several scholars have pointed to a disunity among Haller's beliefs, based on what they see as a disharmony between Haller's scientific and his religious views, in particular between knowledge based on science and truths based on revelation.[3] Yet, although there is evidence, especially in Haller's early poetry, that this issue was of deep concern to him, when one looks at his scientific work this conflict essentially disappears. What one finds instead is that Haller carefully defined for himself the boundaries of scientific explanation so that a clash between scientific knowledge and revelation could not occur. A major factor not just in his debate with Wolff but also in Haller's controversies with Buffon, La Mettrie, Voltaire, and others, was indeed just this issue: science must be carried on within the limits of religion. Where there is a danger of scientific theories forming a basis for materialism and atheism, they must be rejected. As Erich Hintzsche has expressed it, "Haller saw the task of scientific investigation not to arouse doubt but rather to produce confirmation for belief in a Creator" (cited in Toellner 1971:5).

Haller believed that God had created the world in such a manner that only man can understand its harmonious design. "Man alone," Haller wrote, "knows the beauty, the splendor of nature. The wonderful construction of animals, of plants, can raise us through its accordance with their revealed design to knowledge of a Creator: so much we know, it is man alone who knows and enjoys the variety, the order, and the relationships among the parts of the world" (1775–77, 3:56–57). Through human ingenuity and proper methods of study, nature's secrets can be revealed, at least to the extent intended by God. The world is a Divine Creation, and it is through studying its laws that man can come to appreciate the wisdom of the Creator.

Haller's religious beliefs, especially in their relationship to science, were influenced in large measure by his revered teacher Boerhaave. Boerhaave was a "truly Christian" man Haller declared; "to him I owe eternal affection and everlasting gratitude. . . . Perhaps future centuries will produce his equal in genius and learning, but I despair of their producing his equal in character" (1774–77, 1:757). Like his predecessor, Haller sought to combat atheism and materialism wherever he saw them, for he thought that it was only by failing to study the natural world that skeptics and atheists had erroneously been led to view nature as independent from God. Of these misguided individuals, Haller proclaimed, "None has known nature well enough to be able to discover for himself the traces of the Deity that beam forth so abundantly and so luminously in the ends and in the order of created things. Where a Hobbes doubted, a Newton believed; where an Offray [La Mettrie] scoffed, a Boerhaave worshipped" (1772a:6).

As we have seen, Haller's religious beliefs colored his views on embryology both when he was an epigenesist and after he converted back to preformation. Ascribing development to "divine laws" as an epigenesist, Haller saw more and more that the problem of how the embryo becomes organized could be accounted for by no cause other than God. In converting to the preformation theory, Haller preserved God's role in the developmental process. Expressing these sentiments in 1765, Haller remarked: "For hence, indeed, it appears to me certain, that the beautiful structure of animals, so various, that it is always perfectly adapted to the proper and distinct habits and functions and manner of life of each; calculated by rules more perfect than those of human geometry, and most evidently accommodated to foreseen purposes, in the eye, the ear, and the hand, and finally, everywhere; can be ascribed to no cause below the infinite wisdom of the Creator" (1765b, 1803 trans.:434). Each particular generation of a new life is but the result of causes established by God at the Creation.

"Enough, there is a God," Haller proclaimed in one of his poems, "nature shouts it out,/The whole construction of the world shows signs of his hand" (Hirzel 1882:57, lines 325–26).[4] The stars in the heavens never lose their course; animal organisms function according to a Divine plan, with their blood circulating in a perfectly adjusted manner. And finally there is the highest creation: "Man, whose word commands the earth,/Is a composite of pure masterpieces;/In him are united the art and splendor of bodies,/No part of him exists that does not show him to be the master of Creation" (p. 58, lines 347–50).[5] Through science Haller saw man's route to knowledge of God's universe. Science can never lead to atheism and materialism but rather to a deeper appreciation of Divine wisdom and power.

The method of science

"I am persuaded," Haller proclaimed in his famous paper on irritability and sensibility, "that the greatest cause of error [in physiology] has been that most physicians have made use of few experiments, or even none at all, but have substituted analogy instead of experiments" (1752b:115). By "experiment," Haller meant both observation and experimentation; and he recommended anatomical dissection, comparative anatomy, and vivisection as proper experimental procedures. In his own work, Haller relied heavily on anatomical observation, as can be seen in his *Icones anatomicae* (1743–54), where Haller used the technique of injecting colored wax into blood vessel networks, previously developed by Frederik Ruysch, to produce his masterful illustrative plates (see Figure 16). Haller claimed in the opening pages of his paper on irritability and sensibility that his theory was based on six years of observations, many with his student Johann Zimmermann, culminating in a series of 190 experiments on live animals in the preceding year alone. To cite a third example, *Sur la formation du coeur dans le poulet* contains an entire volume of hour-by-hour reports of Haller's observations

Figure 15. Herman Boerhaave, 1668–1738. (Courtesy of the Francis A. Countway Library of Medicine, Harvard Medical School)

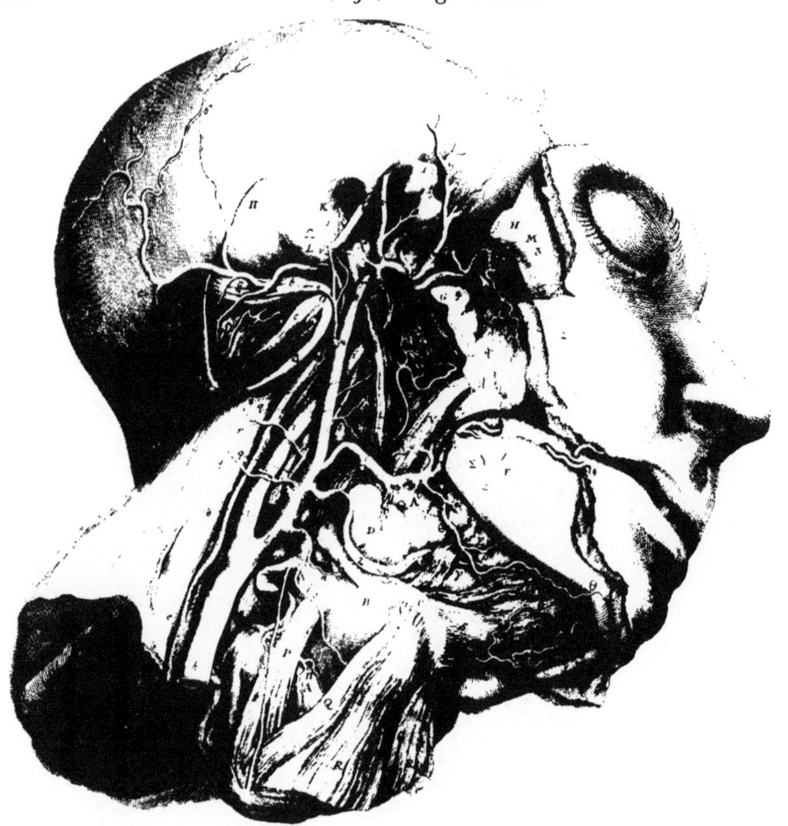

Figure 16. Haller's illustration of the blood vessels of the pharynx. The detail of the arteries and veins was produced by colored wax injections. (From *Icones anatomicae*, 1743–54; courtesy of the Francis A. Countway Library of Medicine, Harvard Medical School)

on hundreds of incubated chicken eggs (which is greatly expanded in the revised Latin edition published in 1767).

Haller cautions against experimenting with prejudices in mind; one must dissect "not with the intention of seeing what a classical author has described, but rather with the desire to see that which nature herself has brought forth" (1757–66, 1:iv). He recommends repeated experiments to sift out extraneous causes, leaving only "the pure things . . . that constantly happen the same way because they flow out of the very nature of the thing" (1757–66, 1:v), and to prevent hasty generalization from isolated facts. "Why do we err?" Haller asks. "We have seen many cases [of something], and we conclude [that it must be so] for all without having seen all" (1772a:45).

Haller was quite clear on the relationship between his own views and the traditions of rationalism and empiricism. "Better telescopes, rounder glass drops [lenses for simple microscopes], more precise divisions of measurement, syringes and scalpels did more for the enlargement of the realm of science," Haller claims, "than the imaginative mind of Descartes, than the father of classification Aristotle, and than the erudite Gassendi. With each step one took nearer to nature, one found the picture unlike that which the philosophers had made of it" (1750a:x). Championing the cause of empiricism, Haller frequently praises Bacon, who "had shown the way to come to know nature through experiment," in contrast to Descartes, who "was too hasty for the experiment, the street was too long for him" (1773:371). And again, in a letter to Charles Bonnet, Haller criticizes Descartes, who "abandoned the routes traced by Bacon and Galileo in order to restore us to the route of arrogant fictions that promises to explain everything without knowing anything!" (Bonnet MSS, 7 November 1773).

Haller's criticisms of Descartes rest principally on the latter's rationalist method. In a preface to the German translation of the first volume of Buffon's *Histoire naturelle*, Haller defends at length the use of hypotheses in science as the most successful route to true conclusions. But one must be cautious in using hypotheses; they must always be tied to experiments or they will lead us astray. This was the downfall of Descartes who "applied a mechanical model to the building and construction of the world, and took the freedom to give to the smallest parts of matter such figures and to impart to it such manners of movement as was necessary for his explanation" (1750a:ix). But, Haller notes, "this convenient custom did not last as long as the idle natural philosophers had wished. The inventions of the imagination are like an artificial metal; it can have the color but never the density and the indestructible solidity that nature gives to gold" (p. x). Arbitrary hypotheses lead to false systems of knowledge; only hypotheses used in conjunction with the proper experimental method are allowable in science.

Haller's rejection of Descartes's method is reflected in his opinions of his German rationalist contemporaries. The reign of scholasticism and metaphysics, Haller wrote in a review of a work on logic, has come to an end in almost all countries. Germany, Haller notes, is an exception, where, led by Christian Wolff, philosophers have taken a step backwards, introducing terms "that had become barbaric" into science, "that domain that Bacon and Galileo had torn away from the schools" (1746b: 356–57). Scholastic rationalism has little to say about the real world, in Haller's opinion, for "God has created individuals, bodies, movements, and one amuses oneself by contriving classes, modes, and qualities" (p. 356).

Haller seems to have rejected rationalism from his earliest student days. In an amusing story, Haller relates an encounter with Cartesianism when he was studying as a young teenager with his stepuncle, Johann Rudolf Neuhaus. "The old man was a determined Cartesian," Haller reports; "he began by making me study the principles of Descartes, and every page revolted me. 'From where do you know that the particles of the second element are round and that those of magnetic material are like a screw?' These questions came again at all moments and drew from me poor response" (Bodemann 1885:89). Haller's negative attitude toward rationalism was undoubtedly reinforced by Boerhaave, Haller's teacher at Leiden, who frequently expressed his own anti-Cartesian attitudes (see Lindeboom 1968:266–68, 270). Thus it was that Haller, a professor at a major German university, was so opposed to the then dominant German philosophical viewpoint. Trained in Switzerland and in Holland, imbued with the principles of empiricism and Newtonianism, Haller rejected both Cartesian and German rationalism in favor of an empirical, Baconian approach (see Toellner 1973). The philosophical gulf between him and Caspar Friedrich Wolff could not have been wider on this point.

Mechanism and forces

"Whoever writes a physiology," Haller declared in the preface to the first volume of his *Elementa physiologiae*, "must explain the inner movements of the animal body, the functions of the organs, the changes of the fluids, and the forces through which life is sustained" (1757–66, 1:i). Haller believed that the science of physiology is the science of movement in living bodies, movement based on mechanical forces. The physiologist, he continues, must explain the forces "through which the forms of things received by the senses are presented to the soul; through which the muscles, which are governed by the commands of the mind, in turn have strength; the forces through which food is changed into such different kinds of juices; and through which, finally, from these liquids both our bodies are preserved and the loss of human generations is replaced by new offspring" (p. i). Sensation, motion, digestion, assimilation, growth, reproduction – these are the functions of the living organism the physiologist must explain. And his explanation must provide, through force mechanics, "a description of those movements by which the animated machine is activated" (p. v). Physiology, Haller proclaims, is "animated anatomy" (1747:5).

Like his teacher Boerhaave before him, Haller endorsed the application of mechanical laws to living processes. Yet this must be done cautiously, for, Haller remarks, "in the animal machine there are many things that are very different from the common mechanical laws" (1757–66, 1:v–vi). Water flowing through a pipe, for example, is not totally analogous to liquids flowing through living vessels, which act in various ways to speed up or slow down the movement of fluids. Simple hydraulics will not explain the motion of the blood through the organism. Haller concludes, however, "I would not for this reason believe in discarding the laws by which moving forces outside of the animal body are governed; I propose that they never be transferred to our animated body machines unless experiments agree" (p. vi).

Haller was a mechanist in his physiological outlook, yet he was not a total reductionist. Rather than reducing vital phenomena to the known laws that govern inorganic bodies, Haller proposed to create a distinct "animal mechanics." Here the basis for explanation was to be laws that operate in the same manner as physical laws, but which are not necessarily the *same* laws. Living organisms thus may possess forces that are not found in nonliving matter; yet these forces operate mechanically in exactly the same way as physical forces do.

As we can recall from Chapter 2, Haller believed that matter is essentially passive and that forces are added to it by God. "Indeed," Haller comments, "the great world bodies move themselves, and thus all parts of the earth and sun with them. But this movement is foreign to them; it is imparted to them. . . . Consequently, movement of matter is not grounded in its being" (1775–77, 1:223). And although motion, and the forces responsible for motion, can be identified through observation, we can never fully understand motion itself. "The measure of forces," he explains, "consists in their effects, for the nature of motion itself, which is a most familiar thing, no one in philosophy has yet comprehended" (1775–77, 1:426). Speaking specifically about irritability, Haller remarks, "One will never know the mechanical source, from which the movements that follow irritation arise, but one will approach [this], one will perhaps succeed in measuring exactly the effect" (1777c:105). All forces operate on a mechanical basis; all can be known through observing and measuring their effects – but their ultimate source is not matter but God. With regard to muscular movement, Haller writes, "We create no movement: our soul wills that the arm lift itself. But it gives to it neither the force nor the movement; God has put the force in the muscle" (1775–77, 3:148).

Haller defended his theory of irritability against the animist Robert Whytt and the materialist La Mettrie.[6] Haller was concerned to show that irritability is a force possessed by animal muscle tissue that is completely independent of the soul. Like gravity and other forces of matter, irritability is a property of a particular kind of matter that is activated by specific stimuli. The source of all of these forces is the same, for, as Haller remarks, "God gave to bodies an attractive force and other forces, which once received are exercised" (1757–66, 7:xii). Furthermore, with God the source of all material forces, Haller argued, they cannot be used to support materialism, as La Mettrie had attempted to do. In his *L'Homme machine* (1748), La Mettrie had used irritability phenomena to bolster his argument against the existence of the soul or any other immaterial substances. Yet, Haller countered, he had shown that irritability is completely separate from the soul; therefore, both must exist (1752b:158).

Along with his eighteenth-century contemporaries, Haller identified materialism with atheism. If there were no spiritual soul and if matter could, on its own, possess active powers, then the need for God as creator and ruler of the universe would be seriously threatened. As we saw earlier, the dangers of materialism and atheism figured significantly in Haller's eventual rejection of epigenesis. No material forces could by themselves produce a living creature out of unorganized matter; otherwise what role would be left for God to play in generation? "Beware," Haller wrote to Bonnet, "that it is very dangerous to admit the formation of a finger by chance. If a finger can form itself, a hand will form itself, and an arm, and a man" (Bonnet MSS, 27 May 1766). Yet atheism and materialism need not be the result of a mechanistic approach to physiological phenomena. The science of animal mechanics can exist, Haller maintained, as yet another proof of the wisdom and design so evident in God's creation.

Haller's Newtonianism

Haller's philosophical beliefs about science bear fundamental similarities to those of Newton, both as expressed by Newton and as seen by his contemporaries. Evidence exists to show that Haller was familiar with the works of Newton and of Newtonian proponents and that Haller held Newton in high esteem as the paragon of proper science.[7] During Haller's years of study at Leiden, the faculty included not only Boerhaave, one of the first Newtonians on the Continent (Lindeboom 1968:268–70), but also 'sGravesande, an early popularizer of Newton (see 'sGravesande 1720–21). After Haller received his medical degree, he traveled to England, where he was enormously impressed with English intellectual life. "In the sciences," he wrote in his diary, "it appears that no land is superior to England" (Hintzsche and Balmer 1971:93; see also pp. 94 and 98). Haller learned to read English after he returned from his journey, and the influence of English styles of poetry is evident in Haller's own poems.[8]

Newton received high praise in several of Haller's poems. In "Gedanken über Vernunft, Aberglauben, und Unglauben" (1729), for example, Haller writes, "A Newton exceeds the limits of created minds,/Finds nature at work and appears as master of the universe;/He weighs the inner force, that is active in bodies,/That makes one fall and moves another in a circle,/ And he breaks open the tables of the eternal laws,/Once made by God and never broken" (Hirzel 1882:46, lines 51–56).[9] Praising Newton's achievements in another poem – the infinitesimal calculus, gravitation, color theory, and Newton's theory of the tides – Haller declares, "He fills the world with clarity/He is a continual source of unrecognized truth" (p. 73, lines 267–68).[10]

That Newton should appear in such a praiseworthy manner in eighteenth-century poetry does not seem all that unusual. Yet, as Richter (1972:66) has pointed out, this is quite unique with regard to the German-speaking world. In fact, Richter claims, Haller was one of the first, if not *the* first, to refer to Newton in his poetry. References to Newton also abound in Haller's scientific writings, where Newton is often represented as the champion of proper scientific method.[11]

It is not surprising, given the scope and volume of Haller's reading, that he should have been acquainted with the major Newtonian publications of his day. Yet their effect on him was far deeper than mere familiarity. In each of the principal aspects of Haller's philosophy of science discussed previously one can trace a clear and fundamental Newtonian influence. In my opinion, Haller consciously sought to emulate the Newtonian program in his scientific work and to construct, in particular, a new physiology based on the canons of the new philosophy.

Concerning the proper method of science, Haller was in close agreement with Newton. Observation and experiment should replace unbridled hypothesizing. Newton's famous declaration that he will "feign no hypotheses" concerning the cause of gravity, because hypotheses not based upon phenomena "have no place in experimental science" (1713, 1934 trans.: 547),[12] finds its echo in Haller's silence concerning the causes of irritability and sensibility. "But the theory," Haller cautions, "why one or the other of these properties is not in these parts, or is in other parts[13] of the human body, such a theory, I say, I certainly do not hope to give. For I am persuaded that the origin of both abilities lies hidden in the intimate fabric, and is placed far beyond the power of the scalpel or the microscope: beyond the scalpel or microscope I do not make many conjectures" (1752b: 115). And even though we do not know the cause of irritability, we can postulate its existence from its observed effects "to which moreover it is unnecessary to assign any cause, just as no probable cause of attraction or gravity is assigned to matter [in general]. It is a physical cause, . . . discovered through experiments, which are evidence enough for demonstrating its existence" (1752b:154). Irritability, itself an attractive force according to Haller, operates in exactly the same manner as gravity.

That gravity, and especially the fact that its origins are unknown, would have been used by Haller as an analogy for his own unexplained force is not particularly unusual. In the eighteenth century, gravity was frequently called upon by physiologists to sanction a myriad of unknown principles and powers that could not be explained but must, it was argued, be postulated to exist (see Hall 1968). Yet in Haller's case, we find a utilization of Newtonian views on matter and forces that exceeds the simple analogy of irritability and gravity. As Newton argued in the *Opticks* (Query 31), matter alone possesses only the passive principle of inertia. To matter forces must be added, for "we meet with very little Motion in the World, besides what is owing to these active Principles" (1730:399). Haller's views on the relationship of matter to forces were, as we have seen with regard to both embryological development and irritability, nearly identical.[14]

Yet if forces must be added to matter, where do they come from? For both Newton and Haller, the answer lay in the religious context of their scientific views. Haller saw God as governing the operations of the world through the forces that He imparted to matter at the Creation. This was also a salient feature of Newtonianism in the eighteenth century. As Heimann (1973:1) has noted: "Newton's ideas were originally presented and disseminated (by the Boyle lectures) in a form which stressed the theological dimension to Newton's philosophy of nature. For early eighteenth century thinkers, Newtonian doctrines of the passivity of matter, of the primacy of forces in nature, and of gravity as a power not essential to but imposed upon matter expressed a theology of nature."[15] For Newton and many of his contemporaries, nature and the laws of nature were seen as dependent upon Divine Providence. An intelligent agent created the world, not blind fate. "Such a wonderful Uniformity in the Planetary System," Newton declares in the *Opticks*, "must be allowed the Effect of Choice. And so must the Uniformity in the Bodies of Animals" (1730:402). Or, as Newton phrases it in the General Scholium to the *Principia*, "Blind metaphysical necessity, which is certainly the same always and everywhere, could produce no variety of things. All that diversity of natural things which we find suited to different times and places could arise from nothing but the ideas and will of a Being necessarily existing" (1713, 1934 trans.: 546). All of the uniformity, diversity, and design in the universe gives testimony to God's authorship.

Haller's beliefs concerning the relationship of God to His creation present a similar viewpoint. Matter possesses no forces or powers except through God's agency. "A first cause," Haller declares, "has thus allotted to different classes of matter abilities and forces calculated according to a general plan, and it is there that we recognize the hand of the Creator" (1751b:86). Blind forces, on their own, could never produce the ordered, yet diverse, world that we know. The constancy we observe testifies to God's will, not chance, as the cause of phenomena. And the diversity is evidence for God's intelligence, not necessity, as the guiding factor. Speaking of the similarities and varieties one finds among flowers, Haller observes, "All is not chance, otherwise the carnation would become a tulip; all is not necessity, otherwise the carnation would remain always such as was the first carnation" (1751b:87). An intelligent God, through His own free choice, created the universe and the laws that govern it.

Haller saw science as leading toward a deeper appreciation of and reverence for God and away from the dangers of atheism and materialism. Newtonian philosophy was also seen as a bulwark against irreligion both by Newton and by his contemporaries. Richard Bentley's Boyle lectures (1693), the final three of which were titled *A Confutation of Atheism from the Origin and Frame of the World*, were designed to make just this point, as were those delivered by other Boyle lecturers. As Roger Cotes declared in his preface to the second edition of the *Principia*, "Newton's distinguished work will be the safest protection against the attacks of atheists, and nowhere more surely than from this quiver can one draw forth missiles against the band of godless men" (Newton 1713, 1934 trans.: xxxiii). Haller saw his own work as adding another arrow to the quiver and as furthering the Newtonian cause of science in support of religion.

In the three major areas of his philosophy of science, Haller was fundamentally inspired by Newton and the Newtonian philosophy. Much of this influence came not from Newton directly but through other Newtonian proponents, especially Herman Boerhaave, whose own example as a scientist and religious man Haller sought to emulate. In Haller's emphasis on experiment, in his views on the relationship of matter to forces, and in the religious context of his mechanistic outlook, Haller expressed a characteristic eighteenth-century philosophy, one derived fundamentally from the Newtonian world view.

WOLFF'S PHILOSOPHY OF SCIENCE

The most formative influence on Wolff's philosophy of science was his education at the University of Halle, which he attended from 1755 to 1759. Founded in 1693, Halle was one of the major intellectual centers of the eighteenth century, its initial faculty including Christian Wolff, Friedrich Hoffmann, and Georg Ernst Stahl. Christian Wolff joined the faculty as professor of mathematics in 1706, only to be banished from Prussia in 1723 at the urging of a group of Pietist theologians whom he had aroused with his philosophical writings. After teaching at

Marburg for several years, Christian Wolff, now quite well known, was recalled to Halle in 1740 by Frederick the Great, and he remained there as chancellor of the university until his death in 1754. The two professors of medicine, Hoffmann and Stahl, taught at Halle from the first years of its inception, Hoffmann remaining there almost continuously until his death in 1742, Stahl until 1715. None of the three was still at the university when Caspar Friedrich Wolff arrived in 1755, yet the context of philosophical and biological ideas fostered by them played an important role in Wolff's intellectual development.

Christian Wolff, and the "Wolffian philosophy," was of enormous influence on Caspar Friedrich Wolff. Imbued with the principles of rationalism, C. F. Wolff set out to create the first rational embryology. But this embryology also had its roots in the mechanism–vitalism controversy that had arisen between Hoffmann and Stahl during the early years of the eighteenth century. Remnants of this debate were still existent at Halle when Wolff began his studies (Gaissinovitch 1956–57; 1961: 214), and his own orientation toward biological explanation in his embryology represents a synthesis of the two views through the principles of rationalism. Never a total mechanist or an absolute vitalist, Wolff attempted instead to steer a middle course between the reductionism of mechanism and the inexplicability of vitalism.

Rationalism

It was Caspar Friedrich Wolff's intention to be the first person to apply the principles of rationalism to embryology and thus to be the first to offer a bona fide explanation for generation. "Since the reasons of the organic body are to be given in the theory of generation," Wolff claims, "this will give philosophical knowledge of it, and therefore it will be defined as the science of natural organic bodies. Furthermore, since anatomy teaches purely historical knowledge of the same subject, the theory of generation will be to anatomy . . . what philosophical knowledge of a thing is to historical knowledge of the same. And therefore one could properly call it *rational anatomy*" (1759:5–6, §§10–11).

Caspar Friedrich Wolff's distinction between philosophical and historical knowledge was taken directly from Christian Wolff, who, in his *Philosophia rationalis* (1728), distinguished among three types of knowledge: historical, philosophical, and

Christian Wolff
Königl. Schwed. u. Hochfürstl. Hessischer Ho
u. Regierungs-Rath, der Mathematick u. Phi
sophie fürnehmster Professor, wie auch Profess
Honorarius zu St. Petersburg, der Königl. Preußisch
Academie, der Königl. Groß-Brittannischen und endlich d
Königl. Preuß. Societät der Wissenschafften Mitglie
Brühl sc. Lipsiæ

mathematical (that is, quantitative) knowledge. "Philosophical knowledge," Christian Wolff declared, "differs from historical knowledge. The latter consists in the bare knowledge of the fact. The former progresses further and exhibits the reason of the fact so that it can be understood why something of this sort could occur" (1728, 1963 trans.:5).[16] The key element that distinguishes historical from philosophical knowledge is the principle of sufficient reason. "Nothing is without a sufficient reason why it is rather than is not," Christian Wolff explained, "that is, if something is posited to be, then something must be posited whence it is understood why this is rather than is not" (cited in Gurr 1959:39). Adopted from Leibniz, the principle of sufficient reason played a key role in the Wolffian philosophy.

One of the most striking aspects of Caspar Friedrich Wolff's writings, particularly the *Theoria generationis*, is their scholastic, deductive style of presentation. Wolff's dissertation is a model of the "mathematical method" championed by Christian Wolff as the universal language to be used by philosophers and scientists alike (see Frängsmyr 1975). The *Theoria generationis* extends this method to embryology through a deductive scheme based on principles, definitions, scholia, and syllogistic reasoning. "By science," proclaimed Christian Wolff, "I mean the habit of demonstrating propositions, i.e., the habit of inferring conclusions by legitimate sequence from certain and immutable principles" (1728, 1963 trans.:17). "One truly explains generation," Caspar Friedrich Wolff echoes, "who deduces the parts of the body and the mode of its composition from known principles and laws" (1759:5, § 5).

The role the principle of sufficient reason played in Caspar Friedrich Wolff's embryology cannot be overestimated. Generation can be explained, Wolff declares, only if one "gives the sufficient reason of the body. And one verifies the theory [of generation] who demonstrates the existence of principles and laws, and shows the sufficient connection between these and the generated body, or proves that if the former is assumed, by necessity the latter is assumed" (1759:5, §§ 6–7). The key to explanation is logical demonstration on the basis of the principle of sufficient reason. Wolff believed that he had reached this level of explanation with his secretion–solidification model for

Figure 17. Christian Wolff, 1679–1754. (From the frontispiece to Ludovici, *Ausführlicher Entwurff einer vollständigen Historie der Wolffischen Philosophie*, 1737)

development and his *vis essentialis*. "The essential force," he declares near the end of his dissertation, "along with the ability of nutrient fluid to solidify, constitutes the sufficient principle of all vegetation [development] both in plants and in animals" (p. 115, § 242).

Wolff believed that before him only descriptions of generation had been given, that no bona fide explanations had ever been offered. "I believe that I do not err very much," he remarks in his *Theorie von der Generation*, "when I say that despite the many works that in ancient as well as recent times have been published on generation, nevertheless up to now no one has given a true system [*Lehre*] of generation . . . or has really explained generation" (1764:2). Wolff, however, makes one concession, for his fellow rationalist Descartes, whose work on generation Wolff claims not to have read when he wrote his dissertation. Although his system was "as false as possible," Wolff asserts, Descartes "showed what an explanation must look like, and he taught how one must philosophize if one wants to do it really and not just to have the appearance of having done it. And in this lies the most distinguished merit of Descartes" (p. 6). What Wolff seems to have admired the most about Descartes's fermentation model of development (see Chapter 1) was Descartes's attempt to show in a deductive manner how the parts of the embryo must necessarily develop on the basis of certain principles about matter and motion. Descartes failed in his effort, yet, Wolff maintains, his is the only theory of generation before Wolff's to offer a real explanation.

Wolff presents an example in the *Theorie von der Generation*, which had also appeared in his dissertation, of a proper philosophical explanation. His topic is the difference between plants and animals, and why animals possess a heart while plants do not (see also Chapter 3). Wolff argues that one can deduce the reasons for this from the observable fact that animal substance solidifies much more slowly than plant material and in fact never reaches the same degree of rigidity as plant substance does. Because of this difference in solidification capabilities, the movement of fluids propelled by the *vis essentialis* through plant substance will be able to form only parallel vessels, while in animal substance, because it is much less quickly rigid, branching vessels will be formed. Branching vessels, however, necessarily entail one common vessel, namely, the heart, from which all others ultimately arise. The possession of a heart, furthermore, determines the differences between plants and animals; from this central determinant, "all the rest of the properties, through which the animal organic body differs from the organic body of a plant with regard to its composition, necessarily must follow." Consequently, Wolff concludes, "if thus I say in this way someone comprehends the building of this organic body from the nature of the forces that form it and thus comprehends it as a result of its causes, then he has a philosophical knowledge of it which is very different from merely historical knowledge" (1764:10).

Wolff believed that scientific method should be based both on logic and on empirical findings. The scientist should aspire to philosophical knowledge, but philosophical knowledge must be based on historical knowledge. Wolff's own repeated anatomical investigations and observations on plants, chicken eggs, and other organisms testify to the importance of empirical data in his research. Yet one must not stop here, for it is through logical reasoning that one can understand why observed phenomena are as they are. Wolff's vision, based as it was on rationalism, was thus of an interconnected, logical universe, accessible to those who follow the proper method of inquiry. And through logic and empirical investigation, one can uncover the rational structure of the world, and one can explain the logical necessity of the phenomena that it exhibits. This is what Wolff believed he had done for the science of embryology.

Mechanism, vitalism, and reductionism

The crux of the conflict that had occupied Hoffmann and Stahl in the early part of the eighteenth century was the extent mechanism could be used in explaining physiological phenomena.[17] Hoffmann argued, in his *Fundamenta medicinae* (1695), that "Medicine is the art of properly utilizing physico-mechanical principles, in order to conserve the health of man or to restore it if lost. . . . Like all of nature, medicine must be mechanical" (1971 trans.:5–6). Stahl believed, on the contrary, that life could not be reduced to mechanical causes; rather the soul (*anima*) controls the vital functions and resists the decomposition the material body would undergo if left to physical forces alone (see Stahl 1708).

C. F. Wolff has often been cited in the secondary literature as a vitalist, principally because of his concept of a *vis essentialis*.

Yet, although not a reductionist, Wolff was wary of vitalism also. He turns to just these issues in his dissertation when he asks, "how are life and the machine mutually connected together in natural organic bodies? Do they accordingly both depend on a common cause, or indeed one upon the other? And if the latter is true, what does life contribute to the machine or the machine to life?" (1759:9, § 36). Wolff's answer is presented in the closing section of the *Theoria generationis*, where he discusses the nature of "mechanical medicine."

Wolff decried the total reductionism he saw in the application of mechanism to biology being made in his day. Mechanical causes, that is, those resulting from the way in which the body is made up of its parts, are accessory, not essential, to vital processes. Consequently, Wolff alleges, "mechanical medicine, whether it exists already developed or may hope to be further refined, is thus far an imaginary system . . . to which there corresponds nothing that exists in the nature of things" (1759:124, § 255). In the process of digestion, for example, chewing and swallowing are mechanical accessory causes. But the assimilation of nourishing fluids is a vegetative process, not based on mechanical causes. "If you now compare," Wolff argues, "the importance of these [mechanical] actions . . . with the ability [of the animal] to maintain itself continually, constantly building and growing . . . will you be able to doubt either, as I have pointed out already, that the machine and whatsoever actions that depend on it are now to be distinguished from the *animal itself*? or indeed, that all these enumerated mechanical actions are to be considered only as an unimportant appendage of animals?" (pp. 125–26, § 255, scholium 1). Because vital activities are not mechanically based, mechanical medicine is not capable of accounting fully for the activities of the organism.

Wolff also rejects mechanical medicine on philosophical grounds, arguing that its proponents have not followed proper explanatory procedures. They have merely pointed to things that occur together and called one the cause and the other the effect. But they have neglected, Wolff charges, to demonstrate the sufficient reason connecting cause and effect in their arguments. Mechanical causes, Wolff admits, are part of vital processes, but they are not their true cause. "Do not think," Wolff cautions, "that the whole matter has been brought to light through a demonstration of the existence of this thing which you take for a cause, and through a deduction of the effect from it; but consider such an explanation as a proposition lacking a demonstration, and add one to it" (p. 127, § 255, scholium 2). The connection between cause and effect, the sufficient reason for the existence of the effect, must be demonstrated. If the proponents of mechanical medicine had followed this procedure, Wolff testifies, they would have realized the folly of their undertaking.

But if life is not to be reduced to mechanical principles, is it to be explained vitalistically? Wolff answers no to this also, although somewhat ambiguously, in his dissertation. First of all, it is clear that Wolff separates the soul from the vegetative process. As he argues,

> what do we mean properly by vital actions? Life is attributed to the animal either insofar as it thinks and moves voluntarily, and thus on account of the acting soul; or for the reason that through various motions occurring in animals, whatever these may be, the maintenance and augmentation of the body are continually accomplished. Both are observed in animals; only the latter is observed in plants. Thus, on the one hand, sensations, voluntary motion, ratiocination, and the remaining [aspects of] thinking, [and] on the other hand, chylification, sanguification, and whatever actions contribute to conservation can be called vital actions, seeing that in both cases life is composed of these. But the former are called animal, the latter are called natural. [1759:120, § 250, scholium 2]

Those vital activities that pertain to the soul in animals (voluntary motion, sensations, thinking) are added to those of the vegetative body. Consequently, the soul is in no way part of the vegetative process itself.

Wolff also distinguishes his views from those of Stahl, who attributed all vital actions to the activities of the soul. In a passage near the end of his dissertation, Wolff declares,

> All those functions of the body that I have denied to be mechanical, I have not explained in any way, inquiring in fact into the connection that exists between the machine and life, but by no means searching further for the causes of this where it has no dealings with the machine. If therefore you should wish to interpret my mind on this, benevolent reader, you could easily err in this. And certainly indeed and especially I would suffer from it [*paterer*],[18] if you should impute to me the opinion of Stahl, or that received from him and slightly altered that Whytt and other more recent scholars have proposed, in which, namely, the functions that occur in our body are attributed to the power of an immaterial soul, whether acting directly and freely, or coerced by the inconvenience inflicted upon it. [1759:130–31, § 255, scholium 4]

Although Wolff agrees with Stahl that vital processes cannot be explained on the basis of mechanical reductionism, he does not concur on the attribution of these functions to the soul. Rather, on Wolff's system these vegetative processes are to be explained through the essential force and the secretion and solidification of fluids.

Wolff's philosophy of biology thus rests on his separation of "vegetative" processes in living organisms from both mechanical processes and those, like sensation and thought, that depend on the soul. Wolff consistently used the term "vegetative" to refer to the activities of nourishment, maintenance, growth, and development in the living organism, that is, the vital functions. Clearly, it is the essential force that is the key to these vegetative processes and thereby the key to Wolff's philosophical views on the nature of life. Denying total reductionism, yet unwilling to ascribe to vitalism either, Wolff sought to create an explanation for life processes that was mechanical in its own right yet also unique to living creatures.

THE PHILOSOPHICAL DEBATE

Although the debate between Haller and Wolff centered around such issues as the development of blood vessels in the area vasculosa, the formation of the heart, and Haller's membrane-continuity proof, the positions each took on these points of contention were intimately tied to their differing philosophical views. Their attitudes toward empiricism and rationalism, mechanism and vitalism, religion and science all played significant roles in their debate. Occasionally, some of these underlying issues surfaced in the debate, yet their presence can be felt throughout the controversy. The two points that received explicit discussion concern the questions of religion and of biological forces. Both of these were closely connected to each person's embryological theory, on the one hand, and to their views on the nature of scientific explanation more generally.

Religion

The issue of God's relationship to his Creation played a major role, as we have seen, in Haller's views on embryological development. His inability to reconcile epigenesis with his religious views was one of the factors that led to his conversion to preformation. Using the idea of preexistence of germs, Haller was able to explain development in mechanical terms while retaining a place for God as the ultimate cause of the production of new life. "If the first rudiment of the fetus is in the mother, if it has been built in the egg, and has been completed to such a point that it needs only to receive nourishment to grow from this, the greatest difficulty in building this most artistic structure from brute matter is solved. In this hypothesis, the Creator himself, for whom nothing is difficult, has built this structure" (1757–66, 8[1766]:143).

Haller must have challenged Wolff on just this point in a letter, for Wolff turns to the relationship between religion and embryological development in a reply to Haller, written on 17 April 1767. Wolff admits that preformation, if true, would be an outstanding proof for the existence of God. Yet, "it is true," he continues, "that nothing is demonstrated against the existence of Divine Power, even if bodies are produced by natural forces and natural causes; for these very forces and causes and nature itself claim an author for themselves, just as much as organic bodies do" (Haller 1773–75, 5:318). Explaining development on the basis of natural causes does not threaten the existence of God, for these natural laws must have been created by God. Besides, Wolff asserts, "by far the clearer and better the proof would be, if, in contemplating the state of nature, we were to find that a single product of it or organic bodies had had need of the Creator, and that nothing organic could have been produced through natural causes" (p. 318).

This is an argument that Wolff expounded in his *Theorie von der Generation* (1764:40–46). Here he asks, why should we admit preexistence of germs for living organisms if we can find no other example in nature of development from causes arranged at the beginning of the world and hidden until their moment of operation?

Wolff objected to the preformationists' reliance on God rather than nature as a cause of generation. "It is of particular importance to me," he remarks at one point in his dissertation, "to discover the principles and universal laws of generation *a posteriori*, and especially to show in any event that the finished plant is not something to whose production natural forces are completely insufficient and which requires the omnipotence of the Creator: inasmuch as this has been observed [in plants], nothing will stand in the way of our allowing the same for the rest of natural organic bodies" (1759:38, § 71, scholium 2). Preformation offers no real explanation for generation; for "those who teach systems of predelineation," Wolff asserts, "do not explain generation but deny that it occurs" (p. 5, § 3). To offer a scientific explanation of embryological development one must show how natural forces and causes are themselves responsible for generation.

In the *Theorie von der Generation*, Wolff laments what our view of nature would be if preformation were indeed the case:

> All organic bodies [would] thus be . . . miracles. Yet how very changed would our conception be of present nature, and how much would it lose of its beauty! Hitherto it was a living nature, which through its own forces produced endless changes. Now it is a work that only appears to produce changes, but that in fact and in essence remains as unchanged as it was built, except that it gradually is more and more used up. Before it was a nature that destroyed itself and that created itself again anew, in order to produce endless changes, and to appear again and again from a new side. Now it is a lifeless mass casting off one piece after another, until the affair comes to an end. [1764:73]

But nature is not this way, Wolff contends, and science must not make it so.

The contrast between Wolff's view of the relationship between God and the world and Haller's conception is indeed striking. Where Haller saw a threat of materialism and atheism if material forces were used to explain development through their actions alone, Wolff saw instead a case of proper scientific explanation based on natural causes. The sterile world of the preformationists was what Wolff rejected. Both believed that embryological development is carried out by means of material forces and natural laws, yet Wolff was able to allow more flexibility, as it were, for their operation in nature. Yet how did his epigenetic theory solve the problems that Haller's had encountered?

Forces and embryological development

Haller objected to Wolff's *vis essentialis* on much the same grounds that he had earlier brought to bear against Buffon's "penetrating force." How, Haller asked, could Wolff's force account for the source of embryonic organization? "Why," he challenges, "does this essential force, which is unique, form parts in the animal that are so different, always in the same place and always according to the same model, if inorganic matter is changeable and capable of taking all sorts of forms?" (1757–66, 8[1766]:117). How could a force be capable of such organizing activities? Haller continues:

> Why does this force always build without any error a chicken out of the unorganized material of a hen, a peacock out of [the material from] a peacock? Nothing is assumed [by Wolff] other than an expanding and progressive force. I would expect nothing more from this than that the net of vessels would necessarily become larger, as long as the expansive force can overcome the resistance. Why, at the site of this net, are formed a heart, a head, a brain, a kidney? Why in each animal is there its proper arrangement of parts? To these questions no response is given. [p. 117]

Wolff's essential force, Haller argues, cannot be the source of organization, because by Wolff's own admission, it is a simple force that operates only through the movement of fluids.

"We do not believe," Haller declares forcefully with regard to all theories similar to Wolff's, "that there is any force that, without being guided by wisdom, can act on matter, following directions that are constantly different and that are properly combined together in such a manner that this brute matter is formed into bones, muscles, organs, and vessels, and so that all of these are joined together in a certain order. All that is produced spontaneously, even those full forms of snowflakes [produced] by artistry, following a single direction, is always formed in hexagons, always in points" (p. 118). Simple formation of snowflakes, crystals, and the like always takes place in the same manner and always produces the same regular structure. But, Haller claims, in the formation of the eye, for example, simple forces could never produce the intricately related membranes in just the right arrangement so that light produces vision. "But these [objections] are infinite . . . ," Haller concludes. "I do not think that this illustrious man has seen anything other than the growth of the chicken, which is guided by heat and the heart" (p. 118)

Wolff argued in his dissertation that the movement of fluids in plants and from the yolk to the embryo in chicken eggs provided observational evidence for the existence of the *vis essentialis*. Furthermore, he distinguished the essential force and the process of solidification from the expansive and resistive forces of Needham, noting that Needham's forces operate on a completely different physical basis than the essential force (1759:106–7, §§ 233–34). Yet beyond this, Wolff does not explain the nature of his essential force any further, concluding simply that "The essential force, along with the ability of nutrient fluid to solidify, constitutes the sufficient principle of all vegetation both in plants and in animals" (p. 115, § 242). The movement of fluids by the essential force, and the solidification of these fluids, produces all of the phenomena of development.

Wolff was not unaware that he had left the nature of the essential force partly unexplained. Yet this was not a major concern for him, as he tells us in the *Theorie von der Generation*:

> It is enough for us to know that it is there, and to recognize it from its effects, as it is demanded purely and simply in order to explain the development of parts. In the name we call it by lies still less; only this must I remind you of, that it is that force through which in the vegetative body all those things are accomplished on the basis of which we attribute life to it; and on this basis I have named it the essential force of those bodies, because, namely, a plant would cease to be a plant if this force were taken from it. In animals it occurs as in plants, and everything that animals have in common with plants depends solely on this force. [1764:160]

The essential force, derived from observational evidence, is that which constitutes life, that which distinguishes the living creature from the nonliving.

Over twenty years later, Wolff published a work devoted to the essential force, *Von der eigenthümlichen und wesentlichen Kraft der vegetabilischen sowohl als auch der animalischen Substanz* (1789). "I flatter myself," Wolff proclaims, "that this essential force, which I laid down indeed at that time [in the early works] as the foundation and also proved, but in no way explained wherein it exists, now . . . through this present treatise will be explained" (1789: 50 n.). Although Wolff's work was written several years after Haller's death, I shall discuss it here in the context of the debate because it sheds light on how Wolff conceived of the essential force both in his early and later works.

Wolff's treatise was written as a response to and published along with two essays on similar topics by Johann Friedrich Blumenbach and Carl Friedrich Born. The Blumenbach and Born papers had been selected for first-place honors in a competition held by the St. Petersburg Academy of Sciences for the best essay on the nature of the "nutritive force." The prize question, originally suggested by Wolff and made public by the Academy in 1782 (see Lukina 1975:414), defined nutrition as the process whereby nourishment is brought to all parts of the body, repair and growth are produced, and the new parts of the embryo are formed. Because this occurs in plants and in animals during embryonic development, the question states, it cannot be due to the heart and must therefore be produced by a force. "Therefore it is asked," the prize question concludes,

> What is the nature of this force? In the first place whether this is the same as the universal attractive force of bodies, or rather, in what way is it seen to be different from this and proper only to living animal substance and to the vegetable substance of plants? If this latter is true, it is asked further, what are its particular effects, and by which properties is it distinguished from the universal attractive force and its singular and special nature made evident? [Blumenbach and Born 1789:ii–iii]

Wolff's treatise, in part a commentary on the essays of Blumenbach and Born, is primarily a presentation of his own answer to this question.

Wolff proposes a very simple model for how the essential force operates. In living organisms, he contends, like substances attract one another, whereas unlike substances repel one another. On the basis of this phenomenon, Wolff claims, one can explain all vegetative activities. In nourishment, for example, liquids are brought to the different parts of the plant or animal, each part attracting out material that is similar to it, which can therefore be used for growth or repair. The attraction is produced by the fact that the two substances are similar in nature and is caused by the presence of the essential force in both the nourishing liquid and the parts of the organism.

In embryological development, attraction and repulsion are responsible for the process of secretion and solidification proposed earlier by Wolff to explain the gradual formation of structures. A solidified part secretes, through repulsion, material that is dissimilar to it. This then solidifies to become a new structure and can grow by attracting material similar to itself from the nourishing liquids. In animals such as the chick, Wolff explains, development commences when the yolk is warmed and begins to dissolve into fluids, which are then repelled from it and are attracted to the site of the new embryo. The result of this movement of fluids is the blood vessel network of the area vasculosa.

Wolff believed that through the attraction and repulsion of the essential force he could explain all vegetative activities, including nourishment, sanguification, embryonic development, and even irritability and sensation in animals. He argues again, as he did in his early works, that mechanical causes (for example, the pumping of the heart after it is formed) influence vegetative activities but do not cause them. One could hypothetically construct, for instance, a model of a plant that was exactly like a real plant in inner and outer structure. But would this model grow and reproduce? "I think this model would remain the same," Wolff answers, "and I think that even the most eager defenders of mechanical medicine would not attribute such activities to the model" (1789:39). The real plant is alive because of the essential force, which "must be peculiar to this plant and animal substance, because no material other than plant and animal substance is nourished, vegetates, or reproduces its kind. And because, moreover," Wolff concludes, "the whole life of plants [and animals], their nutrition, growth, vegetation, and reproduction, rests upon it, one can call it a characteristic and essential force. For where this force is absent, all vegetable processes cease" (p. 39). The structure of the organism cannot be alive without the essential force.

Wolff again distinguishes his force from the vitalism of Stahl, even more explicitly than he had earlier in his dissertation. "This characteristic and essential force," he asserts, "appears to be that, if I do not err, . . . whose existence Stahl very certainly recognized, but which he, incorrectly I think, attributed to the soul. It consists in nothing further than a particularly defined kind of attractive and repulsive force" (1789:42). The essence of life, Wolff contends, need not be attributed to a soul but can, rather, depend upon an attractive and repulsive force.

Wolff further articulates the nature of the essential force in a discussion of Blumenbach's *Bildungstrieb* ("building power"; also called the *nisus formativus*). Here he discusses the kind of objection Haller had raised about Wolff's force, when Haller had questioned how the *vis essentialis* could be responsible for development. Wolff explains here that his own force is not a "building force" and agrees with Haller's position that forces cannot be capable of fashioning, on their own, new organisms. Forces, Wolff argues, are simple in nature and must always produce one invariable effect. Consequently, "the generation or building of the different parts of the body could not depend immediately merely upon one force. For no reason would exist for why this force should work at one time in such a way, at another time in another, or why it should produce at one time one such part, at another time, another part" (1789: 66 n.). If a building force produces, for example, a root at one end of a seed, should it not then have to produce a root at the other end? If building forces were responsible for development, Wolff admits, there would be no reason why development should proceed the way it does. "One must either deny," he concludes, "that nothing in the world can occur without sufficient reason, or one must admit the impossibility of a building force" (p. 67 n.). Either we must give up our belief that everything has a sufficient reason or we must reject the idea of a building force.

But how is Wolff's force different from a building force? The essential force, Wolff explains, is a force that always acts the same in every situation – it attracts when like substances are near one another and repels when unlike substances are near, nothing more. It is the particular *situations* that arise with regard to like or unlike substances being near one another that determine the eventual outcome of the essential force's action. The essential force itself "produces the different parts of the organic body no longer merely through itself and according to its nature, but rather with the help of countless other concurring causes; and what it does through itself alone, becomes a completely simple effect, as attraction or repulsion, and is worlds apart from the building of organic bodies" (1789:67 n.). All of the processes of vegetation are produced by the essential force plus different circumstances. Among these circumstances the chief one is certainly similarity or dissimilarity of substances, which Wolff attributes at one point to "chemical properties," although he does not specify what these may be (p. 53). In order to explain how any particular structure arises in development, one must show how the force of attraction or repulsion has interacted with the existing circumstances to produce the observed phenomena. Thus, the essential force itself is simple and automatically acting, but it produces different ultimate effects in different situations.

Wolff likens the essential force to the universal attractive force, and notes that most phenomena seem to depend on attractive and repulsive forces. Perhaps, he at one point speculates (1789:70), all phenomena are ultimately produced by one single force that is determined differently in different substances (for example, magnetic, electrical, living). Yet, even if this were the case, the essential force would still be a separate force, unique to living matter. It is the same *kind* of force as, for instance, gravity is in its mode of operation, yet it produces the phenomena of life only in plant and animal substance.

Wolff complains in this treatise on the essential force that Haller and others had misunderstood his initial presentations of his theory and had overemphasized the *vis essentialis* to the exclusion of the rest of his ideas. "One could therefore have omitted it and could have attributed the movement of liquids to another cause, as one wanted; or one might have supposed no cause for it, and left the movement unexplained; so this movement of liquids would itself not be denied; and the manner of production and building of parts, as the main point in a theory of generation, would have then still remained the same" (1789: 50 n.). Haller made too much of the force as a name and neglected the theory of development based on the movement of fluids as a total process. Wolff makes a similar point in an unpublished note where he objects to Blumenbach's having compared the *vis essentialis* to his own formative force (*nisus formativus*). "Does this most illustrious gentleman not see then," Wolff queries, "that the motion of humors through a plant is one thing, whereas the formation of a plant is something else? And that therefore the force that moves humors is different from the formative force? Does he not see that by supposing the motion of humors I do not suppose formation, and that by supposing a moving force I do not suppose a formative force?" (Wolff 1973:255). The term "essential force" could even have been eliminated from his theory, Wolff maintains, and his explanation of development on the basis of movement and solidification of fluids would have remained unchanged. "The handle to this calumny, which, however, I endure with good spirit, was given once by the distinguished Haller, who wrote that I derived formation from some force that I called 'essential,' as if the whole matter hinged on this force and the whole explanation of formation consisted of this giving of names. After this, a horde went on to follow this great man" (1973:255).

THE NATURE OF BIOLOGICAL EXPLANATION

My purpose in this chapter has been to contrast the philosophical views of Haller and Wolff and to illustrate the role these differing perspectives played in their debate over embryological development. I have tried to show how each person's philosophical beliefs shaped his view of biological phenomena and dictated support for either preformation or epigenesis. Let me now summarize the major elements of the philosophical split between Haller and Wolff in terms of their differing conceptions of the nature of scientific explanation.

The key parameter in Haller's philosophy of science was certainly his belief in God as creator and ruler of the world. All of his scientific work has reference to this guiding notion, and religious themes are clearly evident in Haller's approval or disapproval of his predecessors and contemporaries. Haller's admiration for Boerhaave and Newton, his comradeship with Bonnet, his criticisms of Buffon and Voltaire, his acrid controversy with La Mettrie, his rejection of Wolff's embryological work, all turn on the theme of reverence for God and condemnation for atheists and materialists.

When we turn to Wolff, we find a far different situation. Wolff was not an atheist; he clearly believed in a divinely created universe. Yet Wolff saw no threat of atheism in science. As he argued in response to Haller, preformation would certainly be an excellent proof of the existence of God, but its opposite, epigenesis, did not lead to a denial of God's existence at all. For where do the laws and forces used to explain generation have their origin if not in God's wisdom? Wolff's belief in a lawlike, logical universe, that could be understood through deductive reasoning, the principle of sufficient reason, and the other tenets of rationalism, allowed him a certain freedom, as it were, from the more confining, more personal God of Haller's beliefs. One is reminded of the Leibniz–Clarke controversy, where Leibniz's more metaphysical God, as creator of a universe based on necessity, is contrasted with the intervening, more closely ruling God of the Newtonian Samuel Clarke (see Clarke 1715).

The differing attitudes of Haller and Wolff on the role of God in the world and on the danger of atheism resulting from scientific theories are reflected in their contrasting views on spontaneous generation. One of Haller's initial reactions to Wolff's theory of epigenesis, expressed in his review of Wolff's dissertation, was that Wolff showed "almost a Needham-like opinion" in ascribing development to an essential force (1760: 1227). In a letter to Haller, Wolff responded to this, claiming "I would hardly think that the opinions of Needham could gain strength from my work, illustrious one, since he dealt with another matter entirely different from mine" (Haller 1773–75, 5:85; letter of 29 December 1761). Needham, Wolff points out, dealt with the generation of animalcules from decaying matter, whereas his own work endeavors to explain normal generation in plants and animals. "But the remaining obscure metaphysical speculations, which Needham adds without any experiments, merit hardly any attention, in my opinion at least" (p. 86).

Haller must have challenged Wolff further on this issue in a letter in response, arguing that even if Wolff's work is different from Needham's, Needham's ideas are consistent with Wolff's and one can, Wolff reports from Haller's letter, "prove them through my [Wolff's] principles as through new grounds" (Wolff 1764:31). Wolff answers, "Against this I can in a fair manner object to nothing. For it is true. And in fact the words cited [from Haller's review and letter] hold nothing further in them than that through my theory, if it is correct, Needham's opinions also receive at the same time a great probability" (p. 31).

Wolff seems almost nonchalant in his admission of a consistency between his and Needham's views on generation. Needham was indeed an epigenesist, and for this Wolff would certainly have approved of Needham's approach. Distinguishing his own specific theories from those of Needham, however, Wolff makes no further effort to condemn Needham's work. This is a far cry from Haller, for whom the mere hint of spontaneous generation was an anathema. That matter could in any sense fashion itself spontaneously into a living creature, however small or primitive, was unthinkable for Haller, for it contradicted all of his fundamental beliefs about the creation of life. Even when he was himself an epigenesist, Haller had expressed grave doubts, in his discussions of Buffon's work on generation, about Needham's ideas. Later, as a preformationist, Haller repeatedly identified Needham's views on spontaneous generation with epigenesis, for both represented the same thing to him; both allowed matter to form living organisms out of unorganized material, completely on its own. In the *Elementa physiologiae*, Haller rejected Needham's views on much the same lines as he did Wolff's. Referring to Needham, Haller charged, "there is in his experiments something that conflicts with my reflections: There is a corporeal force that, alone and without a parent, produces filaments and even spontaneous animals from an inorganic paste.... It appears to us extremely difficult that anything blind and devoid of intelligence could be capable of forming animals according to foreseen purposes, suitably arranged for filling their place in the chain of beings" (1757–66, 8[1766]:112).[19] At no level of simplicity or complexity could Haller allow for the operation of material forces as sole creator of living organisms. "Epigenesis," like its counterpart spontaneous generation, "is totally impossible" (p. 147).

The different attitudes of Haller and Wolff toward religion and spontaneous generation find expression also in their conceptions of biological forces. In all of Haller's discussions of embryological development – in his critique of Buffon's theory, in his rejection of Needham's and of Wolff's – he expressed the same attitude toward material forces. They are the mechanism of development, yet they are not responsible, on their own, for the formation of the new organism.

As presented in 1789, Wolff's views on "building forces" expressed much the same sentiment as Haller's: building forces simply do not exist. Forces must be simple in nature and must operate in the same invariable manner. A building force, Wolff claimed, would have to vary its mode of operation according to variable intended outcomes. Yet this kind of activity would violate the principle of sufficient reason, because there would be no reason why the building force would act in one manner at one time, in another manner at another time. Thus, Wolff agreed with Haller that forces do not, by their actions alone, produce the formation of a new organism.

Although their positions on forces were similar in that they both rejected building forces in favor of simple mechanical forces, Haller and Wolff exhibit differing motivations for their views. The context of each person's conception of forces varied in significant aspects, involving contrasting views on the nature of scientific explanation. For Haller, the overriding concerns in explaining natural phenomena were empiricism, mechanism, and religion. An explanation must be derived *a posteriori* from observational and experimental evidence; if it involves forces, these must be mechanical in operation; and finally, one's explanation must exhibit rather than challenge the role of God as designer and ruler of the world. Haller's Newtonian mechanism and his deeply religious beliefs, also Newtonian in significant respects, joined together to produce a clear conception of how science should explain the phenomena of the world. Haller rejected the materialism of La Mettrie and the animism of Stahl and Robert Whytt on just these grounds. And he rejected epigenesis, in all its guises, from the fermentation model of Descartes, to the spontaneous generation theory of Needham, to the *vis essentialis* of Wolff, as violating the criteria of a proper scientific explanation.

For Wolff, on the other hand, the key aspect of explanation was rationalism. Opposed to mechanical reductionism, to vitalism, and to explanations resorting to Divine Omnipotence, Wolff sought to explain life processes in a manner consistent with his rationalist beliefs. Wolff saw a logical universe, designed by a rational God, that man could come to understand using the proper philosophical method of investigation. Wolff certainly used both observational evidence and mechanical forces in his explanations, yet these were placed in a framework of deductive reasoning, grounded on the principle of sufficient reason. Wolff's was a more *a priori* approach than Haller's, at least more explicitly so, as was the case with his precursors Descartes, Leibniz, and Christian Wolff. He rejected preformation, as he did "mechanical medicine" and Blumenbach's *Bildungstrieb*, all on the basis of their not fulfilling his criteria for proper explanation. And he admired Descartes as having been the only other person to have offered a bona fide, though false, explanation for embryological development. Wolff's quest to create the first "rational anatomy" thus found expression in his epigenetic theory, for through his model of development and the *vis essentialis* Wolff believed he had found the sufficient reason for generation.

Yet I must raise one further question with regard to Wolff's theory. Did his explanation of development, even as presented in 1789, solve the problems that Haller had raised with regard to both Buffon's theory and Wolff's? Did Wolff's final model really account for the source of embryonic organization? We saw that Wolff agreed with Haller's position on forces – that building forces could not exist – and that Wolff defined his own essential force so as to avoid its being thought of as a building force. But if the *vis essentialis* is not responsible for the process of development, what is? What guides the gradual formation of the embryo? Wolff is not explicit in his answer to these questions, for nowhere does he address himself directly to the source-of-organization dilemma. Yet there is an avenue of solution implicit in Wolff's theory, based on his concept of "vegetation" and his proposal, in 1789, that circumstances are responsible for the outcome of the essential force's actions. These ideas are not discussed in any detail by Wolff in his published works, but they are more clearly spelled out in unpublished materials that concern the nature of heredity. In dealing with the question of how traits are passed on to descendants, and of the existence of varieties in the structures of organisms, Wolff reveals his solution to the problem of embryonic organization, a solution that was once again, as we shall see in the following chapter, based on the principles of rationalism.

5
Wolff's later work on variation and heredity

After a lapse of almost ten years in their correspondence, Wolff wrote again to Haller, sending him the latest volume of the Academy of Sciences' journal (1775), in which Wolff had published a paper on the structure of the embryonic heart. In addition to discussing these recent researches, Wolff commented to Haller on his newest project: "The very rich storehouse of monsters that has been collected and preserved over a long series of years in the Imperial museum has now been handed over to me, so that I can compose a description of them and perform anatomies where I decide to. In this therefore it will be necessary to deal once more both with the origin of monsters as well as with generation in general" (letter of 27 September/8 October 1776; see Appendix B, letter VIII). Wolff intended to publish a major work on the subject of monsters, which was to include not only anatomical information, but also theoretical discussions on the origins of monsters, on variation and propagation, and on Wolff's theory of generation. Wolff worked on this project principally during the years 1778 to 1783 (Lukina 1975:415–16), but he never finished his proposed treatise. Those portions that he did complete now exist in the Archives of the Academy of Sciences of the Soviet Union.[1] Predominantly consisting of descriptions and discussions of nine human monstrous specimens (see Rajkov 1964:594–97), Wolff's unpublished materials contain also one lengthier treatise, titled "Objecta meditationum pro theoria monstrorum," which has recently been published both in the original Latin and in Russian translation (Wolff 1973).[2] In this work Wolff discusses not only the origins of monsters, but the nature of variation and heredity more generally.

The subject of monstrous births interested Wolff throughout his career. He included a brief section on monsters at the end of his dissertation (1759:134–35, § 262) and published three separate papers on the subject in the St. Petersburg Academy journal

Figure 18. A chick with four feet. (From Wolff, "De pullo monstroso," 1780)

for the years 1772, 1778, and 1780.[3] To one of these papers Wolff appended a section on the origins of monsters, in which he discussed the work of Duverney, Lémery, Winslow, and Haller, among others, and his own views on how monsters are formed. Lémery and Winslow had carried on a lengthy debate at the Paris Academy of Sciences over whether monsters such as Siamese twins and those with double organs are formed by the accidental union of two embryos during development or are rather one embryo that was designed by God in an abnormal fashion (see Roger 1963:397–418). On either theory, monsters presented a problem for preformation, because, if monsters are formed by accident, then it must be admitted that chance can interfere with God's preordained program for development. And on the other hand, if monsters are preordained, this challenges the wisdom of God, for why would he create malformed organisms, most of which live only a very short while? Haller had in 1739, while in his initial preformationist stage, adopted Winslow's side, arguing that monsters are preformed and that they constitute, in their own special way, further evidence of God's design (see Haller 1739:27–34; Sturm 1974:100–101).

In his paper of 1772, Wolff rejects both positions, claiming simply that "Monsters are not the immediate work of God, but of nature" (1772:567). Wolff challenges the view that monsters are formed by the accidental joining of two embryos on anatomical grounds and, on the basis of his own theory of epigenesis, objects to the idea that monsters are preformed. Referring at one point to a human monster possessing one eye without any optical nerve, Wolff declares, "that the Creator should have delineated on purpose . . . such a structure in the germ cannot be imagined" (p. 562). Only natural causes could have created this disorder. "Monsters are produced by the forces of nature," Wolff concludes, "not evolved from created germs. And if indeed these particular examples of their origin are sufficient, they in turn are at the same time arguments for epigenesis" (pp. 570–71).

In what manner monsters are the product of natural forces, Wolff set out to investigate. Soon he was led to consider the whole problem of variation and its relation to heredity. It is in his work on this question that we find Wolff's views on the relationship between environment and form, the nature of species, and the source of embryonic organization.

"OBJECTA MEDITATIONUM PRO THEORIA MONSTRORUM"

Wolff's principal aim in his manuscript treatise "Objecta meditationum" was to examine the relationship between species and varieties, and to evaluate the effects of environmental influences on the production of both specific and varietal traits. Additionally, he discusses the formation of monsters and the inheritability of monstrous structures such as sexdigitism. Throughout the treatise, Wolff presents a unified view of growth, reproduction, and adaptation based on an expanded explanation of the nature of vegetation. As vegetative bodies, plants and animals possess the ability both to reproduce their species and to respond to changing environmental conditions through the production of varieties.

The nature of species

Wolff opens his treatise by considering the question, what kinds of things can or cannot be passed on in generation from parent to offspring? If a certain structure exists in both a parent and a child, does the parent's structure cause the same structure to reappear in the offspring or is there some other cause that acts in both parent and child? Wolff adopts the latter view, for parents with missing or mutilated parts frequently produce normal children. The principal error of Buffon's system, Wolff points out, was his attempt to explain resemblance by postulating a direct influence of the parents' structures on the formation of the embryo. "Therefore," Wolff claims, "the structure of the parent is not the cause of that structure similar to the parent's that we see return in the embryo; rather, the same, or at any rate a similar, cause has produced both structures, that of the parent and that of the offspring" (Wolff 1973:149, § 9). It is not the parent's structure that produces the embryo's, but rather they both arise from the same cause. This cause Wolff identifies as vegetation: "The forces of vegetation, the act of vegetation, [and] the vegetable substance that now exist reproduced in the seed of the parents, in the egg, and in the male seed, are identical to or at any rate similar to those that once existed [in the parent], by which these [structures] were produced" (p. 149, § 9). Consequently, even if a parent's feet, for example, are amputated, the same cause that originally produced normal feet in the parent will produce them in the child, namely, the process of vegetation. Even if the feet were amputated for a hundred generations, they would still continue to appear in offspring. Only those changes that influence vegetation will be passed on. "Therefore," Wolff concludes, "those things will be able to be propagated that have a sufficient reason in the vegetable substance, whatever their structures or qualities may be" (p. 150, § 9). Structures acquired or altered during an organism's lifetime will not be inherited, Wolff maintains, because these changes do not affect the vegetable substance.

Yet even though the vegetable substance remains constant in reproduction, varieties do arise. In examining how these are produced, Wolff turns to data concerning plants, in particular, to what happens to plants that are transported from one climate to another. Wolff had observed cases of this type in part as a result of seed and specimen collections made by his fellow academician Peter Simon Pallas, who participated in a number of natural history expeditions to different parts of Russia in the late 1760s and early 1770s (see Stresemann 1962; Uschmann 1962). When one transports plants from Siberia to St. Petersburg and vice versa, Wolff reports, one often observes no changes at all for several generations. But then gradually the plants begin to change until they become almost unrecognizable as their Siberian forebears. Yet, Wolff maintains, the species of plants do not change, only their outward structure. Furthermore, he argues, neither the Siberian nor the St. Petersburg form is more "natural" than the other. Both are the product of the plant's species as it is affected by differing environmental conditions.

Wolff bases his claim that the plant's species does not change even though its structure does on the following evidence: Whenever you transfer a plant from one climate to another, he relates, it always changes its structure in such a manner that it will remain distinguishable as a separate species. "Therefore," Wolff maintains, "the species, even if the former figure of St. Petersburg does not remain, nevertheless produces another figure in Siberia that is peculiar to it as its effect, and in this way it distinguishes itself both from the other congeneric species that remain in St. Petersburg and from the effects of these species . . . that have been transferred to Siberia as well – no less than this species distinguishes itself in St. Petersburg from its congeneric species" (Wolff 1973:166, § 32).

Conversely, if a Siberian plant is brought to St. Petersburg and, after a few generations, it changes into a form that is identical to one found normally in St. Petersburg, we should not conclude that one species changed into another. Rather, Wolff argues, we must conclude that we had formerly mistakenly believed that a Siberian and a St. Petersburg plant were separate species, whereas in truth they are simply varieties (1973:167, scholium to § 32). "It is clear therefore," Wolff concludes, "that species, genus, etc., are one thing, whereas external form or structure are another; that species is quite different from this structure or form; that each one of this pair of things has been badly confused in all natural history and also in the natural sciences; that species, whatever it is, is the cause, whereas structure and form are the effect; and that species itself escapes us and our sensations, while only the forms – its effects – are evident and observable" (p. 167, § 33). The species of a plant does not change, even in the production of varieties. Consequently, "species is something in the plant that cannot be observed and that is altogether different from form and structure, because, together with the climate of St. Petersburg, it produces in St. Petersburg the form of St. Petersburg, whereas together with the climate of Siberia it produces in Siberia the form of Siberia, etc." (p. 171, § 35). Wolff thus defined species in such a way that it was totally separate from the inward or outward structure of the organism. This structure is produced by the species (in conjunction with environmental conditions), but there is no aspect of the organism's form that is essential to the species. Consequently, any part of the organism's structure is variable, and no traits are more closely related to, or more expressive of, the organism's species.

Having distinguished species from varieties, and having shown that an organism's species cannot be equated with its structure, Wolff turns to his model of vegetation to explain what the nature of species really is. In the vegetative body, he maintains, one may discern three processes: vegetation, mode of vegetation, and degree of vegetation. As he explains, "Vegetation produces the 'vegetable,' or the plant as such; the mode of vegetation produces orders, genera, species, and classes; the degree of vegetation produces varieties" (1973:168, § 34). Vegetation is the actual process of production of structure in the organism, and it depends primarily on the distribution, secretion, and solidification of humors in the developing organism. The mode of vegetation relates to the qualities vegetable matter possesses and thereby to the organism's species. The degree of vegetation pertains to the size and quantity of structures produced and is affected by changes in living conditions, particularly climate. It is through degree of vegetation that varieties are produced.

The mode of vegetation, which is responsible for the existence of species, depends primarily on what Wolff terms *materia qualificata vegetabilis* ("qualified vegetable matter"). The qualities that vegetable matter possesses, he explains, are "attributes proper to the vegetable substance by which the forces of vegetation are variously determined and which therefore have an essential influence on vegetation" (1973:158, § 18). It is through these qualities of vegetable matter that the organism produces the structures peculiar to its species. "There exist qualified vegetable matters," Wolff claims, "because all matters that vegetate, vegetate in a peculiar and determinate mode so long as they produce a peculiar and determinate organic body through vegetating, and no plant exists that continues to vegetate, first in one way and then in another" (p. 169, § 34, scholium 1). All organisms, even when they change structures after a change in environment, continue to produce that structure proper to those conditions. Never do changes occur arbitrarily. "There exist qualified vegetable matters," Wolff maintains, "which can vegetate but which can each of them vegetate only in its own peculiar mode" (pp. 168–69, § 34).

Wolff points to solidifiability as the most important quality vegetable matter possesses. Differing rates and capabilities of solidification had been used, we can recall, by Wolff in his earlier works to explain such things as why animals have hearts and plants do not. He does not elaborate further in the "Objecta meditationum" on what the different qualities of vegetable matter may be, maintaining only that through the qualifications of vegetable matter, the process of vegetation (that is, the distribution, secretion, and solidification of humors via the essential force) produces different structures in different organisms. "Vegetability is so connected with qualification," Wolff asserts, "that neither unqualified vegetable matter nor qualified nonvegetable matter could exist, . . . perhaps qualification exists in vegetability the way that an attribute exists in a thing" (1973:178, § 36, scholia to scholium 3).

Mode of vegetation, or the qualities of vegetable matter, is responsible for the production of those structures that are proper to the organism's species. But, Wolff reiterates, those structures themselves are not the species; rather, the qualifications of vegetable matter are. As he expresses this:

> Now since we have seen . . . that the mode of vegetation and the qualities or forces of qualified vegetable matters are those causes by which those structures and forms are produced that are usually wrongly taken to be species or genera and orders, it is also clear that those very qualities or forces by which vegetable matters are qualified (or, if you prefer, modes of vegetation) are, properly speaking, those things that we obscurely sense under the name of species, genera, orders and that we are so very well conscious of having observed to be constant. [1973:171, § 36]

It is the *materia qualificata*, or mode of vegetation, that is to be identified with species, not the external form of the organism.

When varieties are produced, as in the case of plants that change their structures in response to a change in climate, the changes in environmental conditions themselves do not cause the change in structure, for they have no influence on the qualified vegetable matter. "It was not therefore the climate or soil, if you want to hit the nail on the head," Wolff claims, "but the plant in its species, its qualified vegetable matter, that, disturbed by the soil and climate, produced a new form and structure by vegetating" (1973:181, § 7). The qualified vegetable matter of the plant remains unchanged when changes in outside conditions occur. It responds by producing new structures that are better able to cope with the new soil or climate. For example, if a plant is moved from rich soil to poor soil or vice versa, changes in the size of its vessels will result, as the plant adjusts to receiving more or less nutriment. Yet this adjustment is not a direct response of the vessels to the soil; rather the mode of vegetation responds to the changed conditions by producing altered vessels.

Those who have identified the structure of organisms with species, Wolff argues, have been in error. "Now therefore it is certainly true," he explains, "that genera and species exist if you understand by species and genera nothing else but what is constant and immutable. This will be false if you place that thing which is constant and immutable in the figure and structure." All aspects of external and internal structure are changeable. "Similarly," Wolff continues, "it is true that varieties alone are mutable if you attribute the whole external figure (and the whole internal structure) to the varieties. But this in turn will likewise be false if after the fashion of the Botanists you understand by varieties certain marks and characteristics in the external form taken apart from other marks in the same form, since whatever is observed in the form ... is equally mutable and able to be propagated" (1973:202–3, § 37). Constancies appear in structures, Wolff asserts, only because environmental conditions are common or remain the same, not because these structures constitute signs of the organism's species. "Therefore," Wolff declares, "the genera and species of the Botanists are reduced to nothing" (p. 203, corollary to § 37).

Even though all structures are potentially alterable on Wolff's theory, we must remember that these changes occur only in a determinate manner, that is, all organisms of the same species change in the same way. New species therefore can never arise from varieties:

Although now the whole organism is mutable and able to be propagated (conditionally able to be propagated, so that it can only be propagated so long as the mutating causes continuously remain the same), it is very clear that true new species or true new genera which are constant can in no way be established for this reason, and that the new structures that are therefore formed by the monster-making potency, however much they depart from the usual structure, are not constantly but only conditionally able to be propagated. [1973:206, §44]

When a variety is produced by the qualified vegetable matter in response to changes in living conditions, that variation will continue in the variety's offspring only as long as the mutating causes remain the same. As soon as these causes cease, the structure of the organism will revert in its offspring to its original structure. Never is a new species created, because the qualified matter itself is not altered by changes in external conditions. The distinction between species and varieties is thus that between the qualified vegetable matter and external living conditions. "Without doubt," Wolff asserts, "the usual structure is a product of the qualified vegetable matter joined with the usual kind of life. Monsters, in contrast, and all other varieties, are products of the same qualified vegetable matter joined with an unusual or monster-making kind of life" (p. 207, § 44).

Thus, no matter what changes are produced in the structures of an organism's offspring, be they varietal or monstrous, none occur equivocally, that is, indeterminately. Furthermore, Wolff reiterates, changed organisms do not alter themselves in such a way that they begin to look like another species. In discussing this point, Wolff makes a statement that has misled some commentators to allege that he believed in the mutability of species, or at least in some kind of limited transformism.[4] Wolff remarks:

Although the mutations that different climates and kinds of nutriments or different degrees of vegetation and quantities introduced into any vegetable body are themselves determinate and peculiar to that body, nevertheless determinate mutations do not for this reason arise so that a new structure of a plant introduced by mutation may return to or at least approach hence some one of the settled species. Rather, as the new structure goes away from and departs from the former structure in varieties and monstrosities defined up to now, so similarly it would go away from and depart from all other species and therefore continuously produce new species, as it were, from the kingdom of possible species, since mutation has gone so far that the former character has been extinguished by it and a new one introduced. And species is in the usual fashion placed in the structure of the plant or indicated by its structure. [1973:234, § 90]

In the final sentence of this discussion, Wolff makes it clear that he is using the word "species" in what he calls the traditional sense of the botanists, that is, to refer to external form and structure. (In fact, in the original manuscript text of this sentence, the word "botanical" is written in, but crossed out, as a modifier of "species.") Thus, when he says that changes in structure "continuously produce new species, as it were, from the kingdom of possible species," he is arguing simply that there are an unlimited number of varieties in structure that can be produced from the realm of possible structures. Therefore, varieties of one species do not approach the form of those of another species. Yet the species of each organism remains constant, because on Wolff's own definition, species is not exhibited by structure.

Wolff's model of development through vegetation thus provided him with a basis for defining species and distinguishing them from varieties, and for formulating a distinction between internal cause and external form that parallels the modern genotype–phenotype dichotomy. Through the way an organism develops and forms its structures, it becomes and remains an identifiable member of a species. Built into this process, however, is an adaptability that allows the organism to produce variations in its offspring in response to changed conditions. Yet the new offspring retains the same qualified vegetable matter and the same characteristic mode of developing that its parent possessed, and thus it remains a member of the same species.

Monsters and hybrids

Wolff distinguishes between monsters and varieties on the basis of the quantity versus the quality of the nutriments. Monsters, he claims, possess structures that are either too large or too small, or they have too many or too few of the same part. "A man may have two heads, but they will be altogether human; he may have three arms, four feet, six fingers, two skulls, three eyes, two noses, but always these parts will be human in a man, bovine in a calf, and of a chicken in a chicken; and there will never be any part in monsters, for example in human monsters . . . that is not a conflation of two portions of two parts, in which portions you recognize the same old ordinary human structure" (1973:229, § 80). This occurs in plants as well as in animals, for in all organisms monstrous structures are produced simply by an unusual quantity of nutriments.

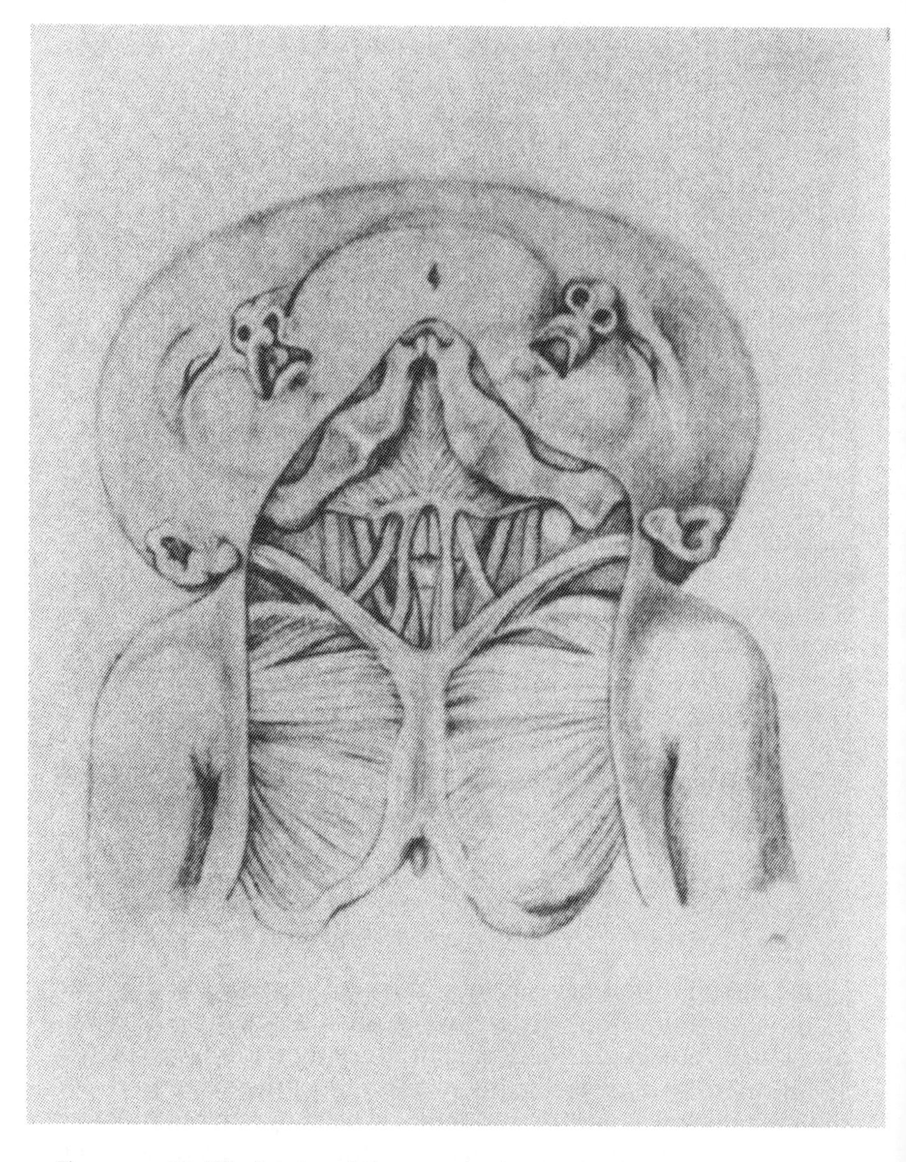

Figure 19. Wolff's drawing of Siamese twins from his manuscript materials. (From Wolff, *Teorija zarozhdenija*, 1950)

Varieties arise, on the other hand, from different qualities of nutriments. In varieties, the organisms's individual parts are the correct number, but they are formed in a different way, be it with a different structure, color, or the like. "An Ethiopian has a black color distributed through all his skin; his bones are yellow, his blood is black. In the usual way there is only one nose, but it is formed differently.[5] . . . [In plants] The Siberian variety has its own peculiar properties distributed through all its parts. In this way there are the varieties of the forests, of the field, and of the gardens" (1973:230, § 82). Furthermore, one and the same organism may possess both monstrous and varietal structures, and they are not always easy to tell apart. Yet their origins are different, for "As all monstrous structure should be ascribed to the degree of vegetation and to the quantity of nutriments proportionate with the digestive force, so in contrast every variety depends upon the mode of vegetation (upon the mode without doubt of production) and upon the quality of the nutriments, or upon the foreign substances that are present in nutriments" (p. 231, § 85).

Wolff's terminology here is somewhat inconsistent with his earlier distinction between mode and degree of vegetation as being that between the production of species and the production of varieties. Now he is ascribing only monsters to the degree of vegetation, and moves varieties to the mode of vegetation. Although this is not consistent usage, we should not be led to any confusion about the distinction between species and varieties. Had Wolff actually polished his treatise for publication, the inconsistency would most likely have been cleared up. Suffice it to say at this point that what he means by the mode of vegetation's being responsible for the production of varieties is simply that it is through qualitative rather than quantitative influences that varieties arise. These qualitative changes in living conditions have no effect on the species of the organism, only "upon the mode without doubt of production." There is certainly room in Wolff's concept of the mode of vegetation for both the unchanging qualified vegetable matter and the mode of production, that is, vegetating that actually gives rise to an organism's structures. By distinguishing this from the degree of vegetation, Wolff sought to underscore his differentiation between monsters and varieties on the basis of the quantity versus the quality of their structures.

Almost all variations and monstrous traits are inheritable only in the sense that if the offspring experiences the same conditions that produced the variation in its parent, it will produce a structure similar to its parent's. If the conditions causing the original mutation change, for example, if the organism is returned to its original climate, it will produce structures that no longer resemble its parent's. These kinds of variations Wolff calls "unprescribed" or "free," and they make up the largest class of variations. Yet there is another group of mutations that are inherited as such, irrespective of external conditions, and these Wolff terms "prescribed." The principal example of this kind of variation is hybridization. When two organisms of different species produce an offspring, that offspring receives a heterogeneous qualified vegetable matter and exhibits structures that are produced by qualified vegetable matter from both parents. Other examples of prescribed variation are, according to Wolff, grafting, where a part of a plant from one species is made to grow from the stalk of another, and the case of an organism being nursed by another, for example, an Ethiopian woman nursing a European child. The plant that results from grafting will be prescribed to produce structures resembling both species, and the child, Wolff speculates, might in some way change its color. Finally, there are prescribed monstrosities, such as sexdigitism, where a monstrous structure is inherited from one of the parents (or even from a grandparent). In all cases of prescribed variation, the new organism's *materia qualificata* is altered. This is the only way the qualified vegetable matter can ever be changed, for in unprescribed variations it remains unaffected.

In the case of hybrid organisms, Wolff admits, because the offspring receives mixed qualified vegetable matter, its species is in that sense changed also. "Since only qualified vegetable matter of vegetable bodies is constant, and since true species consist in this matter alone, it is clear that heterogeneous semen changes both the qualified vegetable matters of vegetable bodies and therefore their true species" (1973:239, § 98). But, Wolff cautions, two things must be kept in mind about this. First, hybrids tend either to be sterile or to revert to one of the parent's species when propagated. "Second," Wolff asserts, "it must be noted that hybrid plants or animals are not new species but old established species that have now been mixed. Therefore, nature does not go beyond prescribed limits in producing and propagating vegetable bodies" (p. 239, § 98). The *materia qualificata* is what provides these limits, what insures that the proper sort of organism results from propagation. Offspring resemble parents because they share similar qualified matter and because they experience similar living conditions; offspring differ from parents because their qualified matter produces altered structures in conjunction with differing environmental influences.

Wolff's species concept and eighteenth-century natural history

Wolff's views on the nature of species, variation, and heredity are of sufficient novelty for us to inquire what relationship they may have had to those of Wolff's contemporaries. First of all, it is clear that Wolff's notions place him squarely in the "essentialist" or typological camp of species concepts (Mayr 1957, 1968, 1969), according to which species are defined by certain essential characters, which may or may not be morphologically expressed, but which remain constant amid accidental variations. Within this group one can also place John Ray, Joseph de Tournefort, Carl Linnaeus, and other eighteenth-century taxonomists, who saw species as constant and varieties as impermanent, and who based their systems of classification upon morphological distinctions between characteristic and accidental structures of plants. For Tournefort, the flower and the fruit were most utilized in classifying plants; whereas for Ray, it was the seed, as well as the petals, calyx, and leaf arrangement (see Tournefort 1694; Ray 1686–1704). For Linnaeus, several different structures were viewed as exhibiting specific characters, the sexual parts of the plant taking precedence as the basis for Linnaeus's binomial nomenclature, the foundation of modern taxonomy (see Linnaeus 1735, 1753; Larson 1971).

Wolff's views on the nature of species present significant differences from those of Linnaeus and the other taxonomists, for Wolff was not particularly interested in classification. Although his interest in plant generation is well established by the amount of discussion he devoted to the subject, particularly in his dissertation, where he formulated the concept of metamorphosis (see Chapter 3), Wolff was not a taxonomist himself. It is difficult to believe that a field naturalist would have formulated a definition of species like Wolff's, which rests on necessarily invisible characteristics. Wolff's separation of species from outward structure, his belief that all external characteristics are variable, and his identification of species with qualified vegetable matter delineate him clearly from his contemporaries. There is no doubt that when Wolff proclaimed "the genera and species of the Botanists are reduced to nothing" he had in mind the taxonomists Tournefort, Ray, Linnaeus, and others. Their systems, relying on outward characteristics of plant structure, can never reveal the true species of the natural world.

A further point must be noted about Wolff's views. In his treatment of hybridization, Wolff allowed for the creation of new species, although, as we saw, he regarded this as only a mixing of established species. Moreover, he pointed out, hybrid organisms tend either to be sterile or to revert to one of the parent types when they reproduce. (This occurs especially in plants that are back-crossed with one of the parent species.) Linnaeus had also been aware of plant hybridization and, in the 1740s and 1750s, hybridization phenomena led him to alter his original view of species as fixed and constant. He began to believe that God had created only the genera and that all species had been formed by crossings between each genus's original species and those of other genera.[6] Although Wolff never went as far as Linnaeus in his views on the significance of hybridization, both clearly recognized that hybrid organisms represent altered species. Wolff was very likely aware of Linnaeus's views; for, even though he does not cite any of the works in which Linnaeus's hybridization theories are expressed, he does refer to other works by Linnaeus in both the *Theoria generationis* and the "Objecta meditationum." Yet Wolff's conclusions concerning the impact of hybridization on the fixity of species were far more conservative than those of Linnaeus, who, in his final theories, saw hybridization as the means whereby the earth had become populated with species. Other botanists adopted positions more similar to Wolff's; Joseph Koelreuter, for example, who dealt extensively with hybridization phenomena, saw no threat in them to the fixity of species. Most hybrids are sterile, Koelreuter noted, and those few that are fertile are not God's but man's artificial creation (see Koelreuter 1761–66; Glass 1959c).

Wolff's emphasis on the influence of climate and other environmental conditions on the structures of plant and animal organisms is also something that was discussed by his contemporaries. Buffon, for example, saw environment as playing a fundamental role in species relationships. Although Buffon's ideas on species changed during his lifetime, he believed, in his final theory, that different living conditions could cause members of the same genus to split into separate species.[7] Thus, for example, the organisms of the New World could be viewed as degenerations from Old World species, the alterations in form having been caused by environmental influences. Wolff quite possibly knew of Buffon's views, even though his only reference to Buffon in the "Objecta meditationum" concerns Buffon's theory of generation. Wolff, like Buffon, allowed environment to play a determining role in the production of plant and animal form. Yet, unlike Buffon, Wolff saw climate as causing only varieties to be produced, not species (although Buffon's distinction between species and variety was never as clear-cut as Wolff's). For Wolff, environmental conditions could never affect the qualified vegetable matter and thus could never alter an organism's species.

Wolff's discussions of the nature of species and varieties show his awareness of other contemporary views, at least those of the botanists, even though his own views were not in agreement. Wolff struck an independent course from plant taxonomists in propounding a theory that was even more "essentialist" than theirs. His own definition of species in terms of qualified vegetable matter owes more to Wolff's rationalist heritage than it does to eighteenth-century taxonomy, for the *materia qualificata* united Wolff's model of vegetation and epigenetic development with his belief in the fixity of species.

Aptness

Wolff includes a rather lengthy section in the middle of the "Objecta meditationum" on the subject of "aptness." Here he explains why organisms are so suitably adapted to their environments and what the origins of adaptation are. If organisms were apt for their living conditions through natural necessity, that is, through a causal mechanism based only on internal and external conditions, Wolff notes, then this would mean that the ends we observe in nature are imaginary. But, he warns, "Beware of precipitating yourself into a most pernicious error! It is by this argument alone, and by no other, that the most manifest ends in nature would also be destroyed, ends that alone, in my opinion, provide us with a solid and firm argument for the existence of God, and by the destruction of which the way would therefore be opened to Atheism" (1973:180, § 2). If the suitabilities we see in the structures of plants and animals were the result of simple efficient causes, rather than a final intelligent cause, Wolff admits, then no longer would the design we witness in nature be evidence of Divine Wisdom.

For something to be "necessarily apt," Wolff argues, the thing for which it is apt must be the cause of its suitability. For example, the formation of teeth, and the arrangement of incisors and molars, would have to be the product, in some sense, of chewing. But this is not the case. Rather, Wolff claims, the wide structure of molars and the sharpness of the incisors are produced by the fact that molars are secreted from a thick arterial trunk, whereas incisors are secreted from the extreme end of the artery. The teeth that form thus necessarily possess these structures. They are also apt for their functions. But they are not neccessarily apt; their formation must be due to an intelligent cause (1973:192, scholium to § 31).

The most convincing evidence for the role of Divine foresight in the formation of organisms, Wolff claims, is their ability to vary their structure when environmental conditions alter. "And now, as I promised," Wolff remarks, "you see a stupendous miracle of Divine Acuity. Not only has an apt fabric been given to plants to correspond to the soil, the climate in which each lives, and the country to which each belongs . . . but also a faculty has been put in these plants of such a nature that if it ever were to happen that the plant would be taken away from its country and transferred into another that is very different . . . it builds a new form and structure for itself." The plant, however, does not lose its species, for if it is returned to its original climate, it soon recovers its former structure. "The same creature," Wolff continues, "as it were, makes itself a new creature for each new country by nevertheless remaining the same as far as its nature is concerned. This seems to me at least to be exceedingly much, and I do not think that I can marvel enough at it" (1773:187, § 15). It is because organisms possess qualified vegetable matter, which allows them to alter internal and external structure in response to changing living conditions, that they show such aptness for their environments. Not only plants, but animals and human beings exhibit this adaptability as well. At one point, Wolff likens the apparent foresight that is universal among organisms to "souls," and he includes human judgment as an apt suitability (pp. 189–90, §§ 21–25).

Wolff's views on adaptability as expressed in the "Objecta meditationum" show us a fuller picture of his attitude toward God and design in nature than is evident in his earlier works.

Yet they are fully consistent with his earlier expressed beliefs, for they further demonstrate his view of a rationally ordered universe, operating on the basis of natural causes. As he stated in one of his letters to Haller, ascribing epigenetic development to natural laws does not threaten the existence of a Creator, because these natural laws themselves must have their origin in God. That all organisms have been allotted the ability to adapt to changing environmental conditions, through a mechanism of natural causes, is further proof of Divine foresight. The organism's qualified vegetable matter, its species, provides the means through which this adaptability, based on an intelligent cause, is expressed. Variation thus serves a purpose in nature, one based ultimately on the wisdom of God.

THE "DISTRIBUTIO OPERIS"

Among Wolff's unpublished papers, there exists a short manuscript that has bearing on the subjects treated in the "Objecta meditationum." Titled "Distributio operis," this document presents a plan for the part and chapter contents of Wolff's major work on monsters. Part 1 was to be an anatomy of monsters that would presumably have included the several anatomical descriptions of monsters that Wolff had already completed. Part 2, titled "Physiology, or on the Nature and Generation of Monsters," was slated to include chapters on the natural orders of monsters, on the generation of animals, on the origins of monsters, on the nature of propagation (where, presumably, the "Objecta meditationum" would have been used), on the souls of monsters, and on the purposes of monsters.[8] A monumental work, Wolff's treatise could easily have stretched into several volumes.

The description of Wolff's chapter on the natural orders of monsters is the most extensive section of the "Distributio operis." The principal question Wolff proposes to examine is whether monsters exhibit the same kind of order that is found in normal organisms. In nature, Wolff asserts, there is a remarkable combination of change with constancy. In monsters, he notes, one observes this as well. Although there seem to be as many different abnormalities as there are different structures, similarities are evident. For example, Wolff observes, one-headed monsters often resemble each other in having two hearts and in having similarly constructed throats, digestive canals, and brains.

"From [this] one must conclude," Wolff maintains, "in monsters of nature, as in the genera of animals and plants, changes combine with preserved similarities in the same remarkable and no less miraculous way" (Gaissinovitch 1961:526–27). And although monsters do not tend to live to an age at which they could reproduce, if they did, their traits would be transmitted to their descendants, which would "become species in future centuries" (p. 527).

Wolff now asks, on what does the observed constancy in nature depend? He answers, as we would expect, that the similarities that species exhibit "have a sufficient reason of their own constancy in the vegetation or vegetation impulse imprinted as if by God in the vegetative substance and which are transmitted to their descendants" (p. 527). No other vegetation can be transmitted to offspring. The purpose of this constancy is thus "the preservation of the species, genera, orders, and the original forms, as intended by God" (p. 528).

Yet Wolff also discusses a second purpose of this constancy: the stabilization of new traits. If monsters could live to a reproductive age, they would transmit their traits to their descendants. Two heads would become "stabilized" as a trait just as six fingers has been. Perhaps one can view monsters as nature's "trials," Wolff asserts; for if a new trait finds access to the vegetative process, then it is transmitted. Sexdigitism is an example of this having occurred, for a six-fingered child may be born even from parents with normal fingers if a grandparent possesses six fingers. The case of national traits among humans, however, Wolff maintains, is not an example of different species having arisen. Rather, the facial characteristics of, for example, Ethiopians, are a result of the effect of climate and soil on vegetation.

"Therefore," Wolff concludes, "there are no necessities" (Gaissinovitch 1961:530). Constancy of form is due to the persistence of vegetation, not to necessary development. "Thus," Wolff declares in opposition to the preformationists, "how unlikely it is that the endless legions of germs in the organic body, already molded and manufactured, could have come from the very hand of God. Apparently, the omnipotent God created only substances that were endowed with their own forces, not apprehensible by our senses and unknowable, becoming apparent only in their activity" (p. 530). Rather than a static progression of predetermined individuals, Wolff views the order of natural things as a product of a dynamic developmental process. God created in organisms the capability of reproducing in their offspring both their traits and their ability to adapt these traits to their living conditions. This takes place through a process of epigenetic development based on the organism's vegetative powers.

I have discussed Wolff's "Distributio operis" in some detail because it has been used by commentators in the past to document Wolff's alleged belief in evolution (Rajkov 1964:610–12). Others have argued that Wolff believed in a limited version of transformism, based on mutability of species within fixed genera (Gaissinovitch 1961:449–50). Yet I do not believe that Wolff's views constitute either evolution or limited transformism. Wolff allows for the transmission of mutations to offspring only when the *materia qualificata*, the process of vegetation that governs the formation of structures, is affected. But this happens in very rare instances – in hybrids and in some kinds of monsters (most of which do not reproduce). Only these kinds of variations are prescribed for the next generation; most varieties and most monsters arise from unprescribed variation, which is the result of the influence of external living conditions on the organism. In hybrids and in those few abnormal traits, like sexdigitism, that are propagated in a prescribed fashion, the altered characteristics of the offspring are indeed the result of heredity. Yet hybrids tend to be sterile or to revert to a parent species when propagated. Furthermore, Wolff maintains, hybrid organisms do not represent new species but rather mixtures of old ones. In the case of monsters, most never reproduce. If they did, Wolff admits, their line could form a new species of organisms. Yet this does not occur; only some abnormal traits are inherited, and these do not constitute new species. Thus, Wolff's views do not lend themselves to a transformist interpretation. He certainly recognized the immense amount of variability that organisms exhibit, in fact more than most of his contemporaries did; yet he retained a place for unchanging species through his concept of qualified vegetable matter.

WOLFF'S VIEWS IN RETROSPECT

Although the material covered in this chapter was written by Wolff long after his debate with Haller and indeed after Haller's death, several aspects of Wolff's theory of variation and heredity shed light on the theory of generation put forward in his earlier works. Furthermore, these later views are consistent with Wolff's philosophy of science, as discussed in Chapter 4, and indeed clarify some of its key issues.

Before attempting a synthesis of Wolff's views, let me treat briefly an issue that has arisen in the secondary literature on Wolff. In a section of another manuscript dealing with a monster consisting of two humans sharing a common chest region, Wolff includes a discussion of the relation of the soul to the body that has led some historians to conclude that Wolff was a materialist.[9] In this passage, Wolff suggests that the soul arises with the body and that it is in some sense the "extract" of the cerebrum and the medullary substance of the body, just as the egg is an extract of the hen and the spinal column is an extract of the egg. (We should recall here that Wolff saw generation as a process of ordered secretion and solidification: the mother secretes the egg in which the spinal column first solidifies; this in turn secretes the limbs and other parts.) The soul is perfected, Wolff maintains, during the life of the body, through its interaction with bodily substances and forces. Yet when the body dies, the soul "throws itself off from the body and continues to live after the body's death, just as we see offspring from other parts of the body live and vegetate after its death. And now eternal life is enjoyed, to which it was destined by the Creator. For our present life is not the true life of the soul, but only its generation and formation; the true life of the soul is this, whereby, liberated from its body it is enjoyed, and which is eternal" (Wolff 1973: 289, §§ 7, 8, 9).

Although Wolff seems to allow for the creation of souls with the formation of each individual, the soul remains separate from the body and is indeed freed from it at the body's death. In some sense the body contributes to the development and perfection of the soul prior to its "true life," which is eternal. I do not see how one can conclude from these passages that Wolff was a materialist, especially if one is careful to use the term as it was used during the eighteenth century. Principally, the epithet "materialist" was applied to those who denied that an immaterial soul coexists in human beings with the material body, and who sought instead to explain such thing as voluntary motion, sensation, and thinking on a material basis alone. "Materialist" was often used synonymously with "atheist," because a denial of a spiritual side to human beings challenged the existence of God. It was principally the antireligious overtones of materialism that led, as we saw in the case of Haller, to its heated rejection. (One has only to recall the outrage that greeted the publication of La Mettrie's *L'Homme machine* in 1748.) Given this contemporary definition, one cannot include Wolff among the materialists. Even though he did allow a close tie between the soul and the body, he never denied that the soul is immaterial; and, furthermore, he allowed it a separate eternal existence, destined for it by its Creator.

We should recall in this regard Wolff's discussion of God and design in the "Objecta meditationum." There he argued that the "aptness" of organisms for their living conditions shows that they were created by an intelligent cause and that the purposes we see exhibited in nature are not imaginary. Wolff believed the universe to be a rationally created product of Divine Wisdom. Never does he suggest that matter alone could be responsible for all life phenomena or that only material substances exist. Wolff did not fear, as the preformationists did, that explaining the generation of living organisms on the basis of natural causes might lead to atheism. Wolff's logical universe could not possibly be the product of material necessity but could result only from a rational God.

As I remarked at the close of Chapter 4, even in Wolff's 1789 publication on the essential force, one problem remained unanswered, the source of embryonic organization. Arguing against the existence of "building forces," Wolff maintained that his essential force was a simple force that acts in the same manner (attraction or repulsion) in all circumstances. Consequently, the essential force can in no way be responsible, on its own, for the formation of the organism. Rather, the differing situations in which it acts result in the development of different structures in the embryo. Yet Wolff did not explain how this process is governed or why it results in the proper organism.

It is in Wolff's unpublished views on variation and heredity that we find his answer to the source-of-organization problem. Through the *materia qualificata*, the determinate qualities that plant and animal substances possess, the transmission of characteristics from parent to offspring is effected. Children resemble parents because they possess similar qualified vegetable matter. And this *materia qualificata* determines the mode of vegetation, that is, the course of development of the organism. It is the qualities of the generative material, and the way they determine the direction of the process of secretion and solidification of structures, that govern generation. Thus, the proper organism results from reproduction, and the species is preserved amid variation.

It was Wolff's rationalist view of substance, which was markedly different from that of many of his contemporaries, that provided him with an avenue of solution to the source-of-organization dilemma. For many eighteenth-century thinkers, matter was a passive entity activated only under the influence of simple mechanical forces added to it by God. This concept of matter underlay Haller's theory, as it did those of most preformationists in the eighteenth century (see Chapter 1). Development had to be programmed in preexisting structures because blind mechanical forces acting on simple, passive matter could not possibly be responsible for producing a living organism. For Wolff, however, whose ideas developed within the tradition of dynamism popularized by Christian Wolff, matter was not such a simple entity. Rather than passive extension, matter was viewed by Wolff as something that possesses form, qualities, modes, and attributes.

Through his view of organism-as-qualified-substance, Wolff was able to give matter a role in generation where his mechanistic contemporaries could not, and thereby to explain why the embryo develops as it does without resorting either to preexisting organization or to self-guiding forces. For Wolff, not all matter is alike; through the qualities it possesses its nature is determined. Applying this to biology leads to Wolff's view of the organism: qualified matter determines the specific nature, the species, of the organism, just as the specific natures of all substances are determined, according to the principles of rationalism, by their inherent qualities.

Still, Wolff's concept of qualified vegetable matter is a problematic one that is not entirely clarified in his writings on the subject. Two things must be kept in mind, however, when considering Wolff's theory. First, it is important to realize that the *materia qualificata* does not refer to an identifiable entity, located at some point in an organism, but rather to a *condition* possessed by living material. Vegetable substance possesses attributes through which it is capable of carrying on the activities of vegetation, that is, through which vital processes are produced. Differing solidification abilities of plant and animal matter are the most important qualities of vegetable matter in Wolff's system. Wolff speculates at one point that perhaps in animals irritability and sensibility may also be fundamental qualifications of animal matter. The qualifications of vegetable matter, in conjunction with the actions of the essential force, thus provide the basis for vegetation, and thereby for life, in plant and animal organisms.

Second, it would be a mistake to conclude that, in his final theory, Wolff resorted to preformation by allowing for some sort of material substrate to be passed on from parent to offspring. Wolff saw his concept of "qualified vegetable matter" as providing the completion of his epigenetic system. Through it, he could explain how the embryo develops from its own forces without resorting to preexisting structures. Admittedly, embryos on Wolff's theory do not start out in a state of absolute homogeneity. Yet one must be careful not to define epigenesis so narrowly that clearly epigenetic systems like Wolff's are excluded. Gradual development of complex heterogeneity from simple heterogeneity can provide a valid epigentic viewpoint (see Churchill 1970a: 169–71). In Wolff's system, the embryo's initial heterogeneity is of a potential nature, based only on physical factors like solidification and attraction and repulsion, which produce the structures of the organism through a gradual, but automatic, sequence of events. This is a far cry from preformation, especially in its eighteenth-century *emboîtement* form.

Wolff's theory of epigenesis, and the philosophy of science upon which it was based, was uncommon for his era. Opposed to mechanical reductionism, to vitalism, to materialism, and to explanations relying only on Divine Omnipotence, Wolff sought to explain life processes in a manner consistent with his rationalist beliefs. Wolff's theories were based on mechanical principles, with their emphasis on secretion, solidification, attraction, repulsion, and the like. But Wolff's mechanism was not based on passive particles in motion, to which life processes were to be reduced. Rather, Wolff's rationalist view of substance allowed for a more complex sort of mechanism, as such a view had for Leibniz and for Christian Wolff, and thereby opened up avenues of explanation that were closed to so many of Wolff's contemporaries.[10] By starting from a different philosophical base, Wolff's "rational anatomy" provided a unique challenge to eighteenth-century preformationism.

6

Epilogue: the old and the new

The principal thesis I have argued in this book is that the debate between Haller and Wolff over embryological development can be fully understood only when one considers the philosophical presuppositions that underlay their controversy. Haller and Wolff debated not only the formation of embryonic structures, but, more fundamentally, the views of scientific explanation upon which their embryological theories were based. Each person brought to bear on the embryological points of contention a whole host of philosophical assumptions concerning mechanical explanation, biological forces, logical reasoning, God's relationship to natural phenomena, and the like – assumptions that fundamentally colored their respective perceptions of the embryological level of debate. At issue was more than simply the interpretation of specific observations of chick development, for it was Haller's whole Newtonian outlook, with its religious and mechanistic parameters, that confronted Wolff's rationalist, antireductionist beliefs. This is what led to such a heated debate between the two, and what was responsible for their controversy's inconclusiveness and its frequent aura of noncommunication.

My thesis is not one that is unique to the Haller–Wolff debate. There is no doubt that major controversies in science regularly have their extrascientific components, be they philosophical, religious, or political in origin. Furthermore, it is these underlying, contextual elements that provide the key motivational forces behind scientific debates. In the history of biology, for example, there have been numerous such controversies, where observational, "scientific" issues provided the forum for clashes between whole metaphysical systems. The debate between Harvey and Descartes over the nature of the heartbeat is one such example, for Descartes's reduction of the heart's muscular contraction to a combustion model opposed far more than Harvey's "pulsific faculty." In contention were two systems of biological explanation and two philosophical approaches to life phenomena. Similarly, one can view the repeated controversies over spontaneous generation, such as Needham's debate with Spallanzani or Pasteur's confrontation with Pouchet; the Stahl–Hoffmann controversy over mechanism and animism; the Lawrence–Abernethy debate over the existence of vital principles; or the clash between Roux and Driesch over embryonic differentiation. Each of these confrontations, as historians of science have increasingly come to realize, rests not only on observational or experimental issues, but on broader philosophical or sociological elements more generally.[1]

A further aspect of scientific controversy that is exemplified in the Haller–Wolff debate is the tendency for such debates to remain unresolved. The myth of the *experimentum crucis* manifests itself here, for most controversies simply are not resolvable in terms of experimental evidence. Can one imagine, in retrospect, a single experiment or observation that would have settled the Haller–Wolff debate once and for all, that would have convinced either Haller or Wolff of the truth of his opponent's position? I think not, for repeatedly in their controversy Haller and Wolff confronted each other with new, seemingly conclusive evidence, only to find the opponent subsuming it under his own explanatory system.

Closely related to the unresolvable nature of the Haller–Wolff debate is the element of incommensurability that is also evident.[2] At a number of points in the controversy, one has the clear feeling that Haller and Wolff really are "talking past" one another, that neither is able to see the same issue, or the same evidence, fully from the other's point of view. Although I do not wish to argue that Haller's and Wolff's positions were *totally* incommensurable (or that such a thing ever exists in science), in a very real sense Haller and Wolff *were* "living in different worlds," as each sought to promote his own conception of how science ought, and ought not, to explain biological phenomena. I do not think, for example, that Haller ever showed any understanding of Wolff's rationalist approach, of his driving need to subject empirical evidence to logical analysis and to demonstrate deductively how an explanation accounts for the facts. Neither did Wolff ever share Haller's deeply ingrained distrust of any scientific theory that might threaten the omnipotence of God. Because of this, Haller and Wolff were never really able to see the empirical evidence from exactly the same perspective, for neither ever fully grasped the implications that such evidence held for the other's philosophical position. This is not to say that there was *no* communication or *no* common ground of understanding between the two. Rather, their debate exhibits that limited degree of communication and understanding that is necessary for controversy to arise and flourish in the first place.

Both the unresolvable and incommensurable aspects of the Haller–Wolff debate are closely tied to my general thesis, for it was because their controversy took place in a much wider context than that of embryological observation alone that it was so inconclusive. Haller and Wolff were not merely trying to convince one another of the "correct" explanation of blood vessel development; they were in fact out to discredit each other's philosophical view of scientific explanation. That neither succeeded in winning their debate is less significant than the fact that neither ever really could, because the rules of the game were so different for each. The Haller–Wolff controversy, like so many others in science, thus provides a "laboratory view" of the relationships between metaphysical commitments, theoretical beliefs, and empirical observation as these operate in the practice of science.

THE RISE OF TELEOLOGICAL EPIGENESIS

Although the debate between Haller and Wolff was not resolved by the two of them, history shows that after the great preformationist triumvirate of Haller, Bonnet, and Spallanzani, the theory of preexistence fell into demise and epigenesis replaced it as the most widely held view of embryological development. In the late eighteenth and early nineteenth centuries, the most important contributions to embryology came from Germany, where Blumenbach, Kielmeyer, Döllinger, Oken, Pander, von Baer, and others pursued developmental researches from an epigenetic point of view. Yet none of these investigators ever really disproved preformation in any significant experimental way; most simply rejected it out of hand. As Oken declared in 1810, "The theory of preformation contradicts the laws of natural development" (1809–11:28). That the German epigenesists were able to dismiss preformation in such a manner points to a deeper cause than the evidence itself, for the philosophical foundations of their biological views were markedly different from those of their eighteenth-century preformationist predecessors.

In the eighteenth century, most embryologists adopted preformation because it was the only mechanistic explanation of development that was consistent with the dominant religious world view. If one believes that matter is essentially passive, activated only through divinely endowed mechanical forces, then preformation is really the only viable explanation for development one can adopt. Such a viewpoint underlay the arguments of preformationists from Malebranche through Haller and Bonnet. Those who proposed mechanistic epigenesis – like Descartes, Maupertuis, or Buffon – either failed to account satisfactorily for the organizational aspects of development or, like Maupertuis, turned to nonmechanistic solutions to obviate this difficulty. As I have argued in previous chapters, it was the source-of-organization problem upon which mechanistic epigenesis continually foundered.

With Blumenbach and the other German embryologists, the situation changed radically. No longer was the source of embryonic organization the central problem to be explained. Rather, organization became the one element of life that was taken for granted. Following Kant, these German embryologists uniformly embraced a teleological view of embryological development, based on a presupposed original state of organization in the generative material.[3] Agreeing with Haller and the preformationists on one count, Blumenbach and his followers rejected the idea that organization could be explained in mechanical terms. As Kant argued, "Absolutely no human reason . . . can hope to understand the production of even a blade of grass by mere mechanical causes" (1790, 1966 trans.:258, § 77). Referring specifically to Blumenbach and his *nisus formativus*, Kant remarked similarly, "That crude matter should have originally formed itself according to mechanical laws, that life should have sprung from the nature of what is lifeless, that matter should have been able to dispose itself into the form of a self-maintaining purposiveness – this he rightly declares to be contradictory to reason" (p. 274, § 81). Once one accepts organization as a teleological fact, Kant proposed, one can then proceed to explain on a mechanical basis how this organization functions and is maintained. Using the concept of purposive organization as a regulative idea, as a guiding thread, one should then conduct empirical investigations into the mechanisms that govern vital phenomena.

This is indeed what the German embryologists proceeded to do, with important descriptive and comparative embryological studies as a result. Freed of the need to explain the source of organization, they rejected eighteenth-century preexistence theories as contradictory to the developmental laws of nature and turned their attention instead to describing the sequence of embryological development in accurate detail. One of the culminations of this research was von Baer's enunciation of the germ layer theory in 1828 (see von Baer 1828–37). Yet one must realize that these embryologists substantially altered the ground rules from those on which their eighteenth-century predecessors had based their embryological theories. Indeed, Kant termed this new form of epigenesis a "generic preformation," built as it was on "preformed" purposiveness (1790, 1966 trans.: 272, § 81). The epigenetic viewpoint proposed by Blumenbach and later German embryologists did not fulfill the criteria of seventeenth- and eighteenth-century epigenesis, for it *relied on*, rather than *explained*, embryonic organization. As Temkin has remarked in a similar vein, "the theory of epigenesis as advocated in Germany around 1800, contained a strong teleological element and was, therefore, a compromise between preformation (rejected in the form of *emboîtement*) and mere mechanistic explanation" (1950:230, n. 16). Opposed to preexistence, these embryologists retained the most successful aspect of eighteenth-century preformation – its assumption of built-in organization – while at the same time promoting a far different view of gradual development than that of their mechanistic forebears.

With the German epigenesists, ontogenetic development also took on wider significance than it had during the preceding century. Rejecting the search for a causal account of generation, these embryologists viewed the formation of the individual as part of a process of development encompassing the whole of the organic realm. As Gasking has pointed out, "Growth accompanied by change was now regarded as a fundamental feature of the Universe, and the growth of living things was the analogy in terms of which all other processes were to be understood. It, therefore, seemed a basic phenomenon requiring no further explanation. What had been an atypical, almost miraculous process from the seventeenth and eighteenth century point of view, became the paradigm of the natural for the [early nineteenth-century] nature philosophers" (1967:151). One consequence of this new "developmental paradigm" was the significance that embryological data began to provide for natural history. A new view of the relationships among species, and of their historical development, arose with the concept of the archetype – the idea that all animal forms are simply variations of a limited number of ideal types – and with the first enunciations of the biogenetic law by Kielmeyer, Oken, Meckel, and others (see Gould 1977). Not only was individual development seen as the result of an immanent teleological power, but so too was the history of life on earth. As Kielmeyer explained, "Since the distribution of forces in the series of organized beings [organisms] follows the same order as their distribution in the developmental stages of given individuals, it follows that the force by which the production of the latter occurs, namely the reproductive force, corresponds in its laws with the force by which the series of different organized beings of the earth were called into existence" (1793:262). That ontogeny recapitulates phylogeny seemed a natural consequence of the belief that the creation of both the individual and the species are under the guidance of immanent, nonmechanical laws of nature.[4]

Such an explicit tie between embryological development and natural history was absent in the eighteenth century. Although Bonnet had drawn a parallel between ontogeny and the perfection of species on earth (Gould 1977), and Buffon, in later volumes of the *Histoire naturelle*, had sought to establish a connection between the history of life and environmental influences on the *moule intérieur* (Farber 1972), the identification of ontogenetic and phylogenetic development by German embryologists and naturalists was based on a far different conception of nature than had existed in the preceding century. Embryological research took on a new significance and was used in answering new varieties of questions. Rather than seeking a mechanistic, causal account of the formation of the individual, the German epigenesists turned to embryological investigation to provide examples of nature's developmental tendencies, as well as data for understanding the taxonomic relationships among organisms. Descriptive and comparative embryological studies were pursued with renewed effort and with remarkably fruitful results. Not until the end of the nineteenth century did descriptive embryology give way once again to causal embryology, with the experimental work of His, Roux, Driesch, and others, and, as a consequence, to a new round of debates over preformation and epigenesis.[5]

In the shift that I have described from eighteenth-century mechanistic preformation to nineteenth-century teleological epigenesis, what role can be ascribed to Caspar Friedrich Wolff? Let me state at the outset that nowhere does Wolff express any of the explicit teleological viewpoint found in Blumenbach and the other Germans. We should recall that Wolff clearly distinguished his *vis essentialis* from Blumenbach's *Bildungstrieb*, rejecting the latter's "building force" on philosophical grounds. Furthermore, Wolff was none too happy with those who had mistaken his essential force for a "building force." Wolff's theory of epigenesis was based neither on the assumed teleological nature of the organism nor on the presupposition of its inherent organization.

That Wolff's theory of epigenesis was different from those proposed by his epigenetic successors was recognized by von Baer, who was familiar with almost all of Wolff's writings (see von Baer 1866:294–96). Against preformation, von Baer wrote, Wolff "fought a good deal and triumphantly . . . and set up against it the principle of epigenesis, of the actual new building of all parts and of the whole embryo. In this he evidently went too far. It is true that neither the head, the feet, or any single part whatever was there earlier, rather all become; but they become not through actual new building, rather through the reorganization of what already exists" (1866:318–19; see also Oppenheimer 1967:295–307). Reproduction, von Baer explained, is a continual process of organization being passed on from parent to offspring.

Although Wolff's papers on the formation of the intestines were relatively unknown among German embryologists until J. F. Meckel's translation of 1812, Wolff's support for epigenesis in his earlier works was undoubtedly much more widely known.[6] Certainly those who read Blumenbach's works were aware of Wolff, although they would not necessarily have been led thereby to read Wolff in the original. Yet the later German embryologists did not explicitly build their theories directly on Wolff's. To be sure, Wolff was regularly cited, especially after 1812, as a precursor to the epigenetic theory. Yet Wolff and his work were out of the mainstream of German biology, because Wolff was at St. Petersburg for almost all of his active career and trained no students there. Ironically, Blumenbach began his studies at Haller's former stronghold, the University of Göttingen, and he initially accepted Haller's preexistence theory before converting to epigenesis (see Lenoir 1980). Thus, although Wolff's views were known among his German successors, his influence was less direct than one might initially suspect.

Even though Wolff was not the sole, or even the primary, cause of the shift toward epigenesis in German embryology, one can identify elements of similarity in his and the later epigenesists' theories. Although Wolff did not endorse a teleological view of the organism, he was much less explicitly concerned about the source-of-organization question than his preformationist contemporaries had been. In his early works, he did not even deal with this problem (much less solve it), leading to Haller's justified criticisms in this regard. In his unpublished manuscripts, I have argued, one can find an account of how organization is passed on through generations, via qualified vegetable matter. Yet even here, Wolff did not offer this as a direct response to the organization dilemma, but rather as an explanation for the fixity of species.

Through his theory of qualified vegetable matter, Wolff explained both hereditary resemblance and variation. And even though his theory represents neither a lapse into preformation nor a teleological view of the organism, it does allow organization to be in some sense passed on rather than created *de novo* with each new instance of generation. Wolff's unpublished work was, of course, completely unknown to his immediate successors, and it therefore had no influence on the German embryologists' subsequent endorsement of organization as an assumed property of living organisms. Yet it is significant to note that neither Wolff nor his epigenesist successors "solved" the source-of-organization problem in the terms in which it was enunciated by seventeenth- and eighteenth-century mechanists. Wolff's rationalist program allowed him to redefine the question in such a manner that his epigenetic explanation based on qualified vegetable matter accounted for the hereditary phenomena of living organisms. Just so, the early nineteenth-century epigenesists viewed embryonic organization not as a problem to be explained, but as a starting point for embryological research.

The rise of teleological epigenesis was not an isolated event in late-eighteenth-century Germany. Rather, it was part of a much broader shift toward developmental and historical thinking in many areas of German intellectual life. Several scholars have pointed to *Naturphilosophie* and Romanticism more generally as providing a philosophical climate conducive to teleological thinking in biology, and to the reciprocal influence of biological theory on these philosophical movements themselves. Others stress the contemporaneous rise of progressivist views of human history, and the social and political events experienced by

German intellectuals in the 1780s and 1790s.[7] All of these analyses point to a fundamental change in philosophical viewpoint that encompassed far more than the biological sciences. Although it is beyond the scope of the present study to explore these issues further, there is ample evidence to indicate that German embryologists turned to epigenesis in the late eighteenth and early nineteenth centuries because it filled the same kind of philosophical need that preformation had during the previous period. Epigenesis was as compatible with the new progressivist view of human history and natural phenomena as preformation had been with the religious and mechanistic beliefs of the seventeenth and eighteenth centuries.

All of this points once again to my principal message. One cannot treat the embryological views held during these earlier centuries in isolation from their wider philosophical context. For Malebranche and Descartes in the seventeenth century, for Haller, Bonnet, Buffon, Wolff, and Blumenbach in the eighteenth, as well as for Kielmeyer, von Baer, and the other German epigenesists in the early nineteenth century, embryological issues were not all that was at stake. The nature of matter and of forces, the applicability of mechanism to biology, the roles of the empirical and the logical in scientific explanation, the reducibility of life to nonlife – these and similar questions formed the background of embryological research and debate. As a forum for the confrontation of these more fundamental issues, the science of embryology thus provides a particularly revealing guide to the intellectual forces that shaped Enlightenment biology more generally.

APPENDIX A

Chronology of the Haller–Wolff debate

1758	Haller	*Sur la formation du coeur dans le poulet*
1759	Wolff	*Theoria generationis*
1759	Wolff	Letter I (23 December)
[1760]*	Haller	Response to Wolff Letter I
1760	Haller	Review of Wolff's *Theoria generationis*
1761	Wolff	Letter II (29 December)
[1762–64]	Haller	Response to Wolff Letter II
1764	Wolff	*Theorie von der Generation*
1764	Wolff	Letter III (20 December)
[1765]	Haller	Response to Wolff Letter III
1765	Wolff	Letter IV (5 May)
1765	Haller	Review of Wolff's *Theorie von der Generation*
[1765]	Haller	Response to Wolff Letter IV
1765	Wolff	Letter V (16 November)
1766	Haller	*Elementa physiologiae corporis humani*, vol. 8.
1766	Wolff	Letter VI (6 October)
[1766–67]	Haller	Response to Wolff Letter VI
1767	Haller	*Opera minora*, vol. 2
1767	Wolff	Letter VII (17 April)
1767	Wolff	Moves to St. Petersburg
1768–69	Wolff	"De formatione intestinorum"
1770–71	Haller	Reviews of Wolff's "De formatione intestinorum"
1774	Wolff	*Theoria generationis* (2nd ed.)
1776	Wolff	Letter VIII (27 September/8 October)
[1777]	Haller	Response to Wolff Letter VIII
1777	Wolff	Letter IX (7 May)
1777	Haller	Death (12 December)
1778–83	Wolff	Unpublished treatise on monsters
1789	Wolff	*Von der eigenthümlichen und wesentlichen Kraft*
1794	Wolff	Death (22 February)

*Because Haller's letters to Wolff no longer exist, the dates given here for them are estimated.

APPENDIX B

Wolff's letters to Haller

Wolff sent a total of nine letters to Haller, seven during the years 1759 to 1767 and the final two in 1776 and 1777. The originals of the first seven and the ninth letter are in the Burgerbibliothek in Bern ("Briefe an Albrecht v. Haller," MSS Hist. Helv. XVIII), while the eighth letter is in the archives of the Germanisches Nationalmuseum in Nuremberg. Haller published the first seven of Wolff's letters in volumes four and five of his *Epistolarum ab eruditis viris* (1773–75). These have also been published, with letter nine, in German translation by Schuster (1941) and in Russian by Gaissinovitch (1961:510–25). All nine letters are included below in full, letter eight being published here for the first time.

LETTER I

Berlin
23 December 1759

I am bold enough to send you my inaugural dissertation [*Theoria generationis*, 1759], which sets forth my theory of generation, even though I hold you to be the highest and most acute judge of these matters; nevertheless your singular humanity, with which you support the labors of those who at least make a sincere attempt to accomplish something, bids me hope that you will regard my efforts favorably and fairly as well.

Nor, illustrious one, does it frighten me that you have recently honored the opinion opposed to my theory with your illustrious name and authority. For not only have I long been convinced of the sincere love with which you pursue the truth, which I believe I have touched upon, and which the whole learned world admires in you, but those things that have been advanced in the most welcome experiments undertaken on incubated eggs are not of such a kind that I ought to have abandoned all hope of the truth of my theory, which, at the time that I first read about the experiments, was almost finished. This is what your words surely say, illustrious one, in Part II, p. 172 [of *Sur la formation du coeur dans le poulet*, 1758a]: "I propose the contrary opinion, which is *beginning* to *appear* to me to be the most probable. The chicken has furnished me with reasons in favor of development [preformation]." What then prevents me, having completed my little work, from submitting it now to your most penetrating judgment as reasons fighting for the other side?

If, illustrious one, I could somehow please you with this work of mine, of whatever quality it may be, I would consider myself sufficiently happy and would then also be confident of the approval of others. For your most solid judgment has been for me up to now and will always be for me equivalent to that of all men. If I have erred or not satisfied your nature, do not deem me unworthy of your correction, however it pleases you to provide it; rather, I would earnestly ask that you allow me to enjoy the same favor with which you are accustomed to regard all your disciples.

I will be most devoted to you all my life.

LETTER II

Berlin
29 December 1761

After I had received your letter, which was full of the highest humanity and benevolence toward me, and the works of the most famous Sproegel,[1] I hesitated in doubt about whether I should respond immediately in order to attest to my most grateful feelings or rather whether I should not disturb you in matters of great importance too often without need but should wait for the publication of your opinion and then fulfill all my duties at once. This latter argument finally convinced me. Now in the very celebrated Göttingen review, which reached me very late through the fault of the Berlin distributor, I have just now read, although they appeared last year, judgments and opinions concerning my dissertation [1759] that I easily attribute to you yourself, illustrious one [see Haller 1760]. Wherefore, I give you my greatest and most obliging thanks, both for the benevolence with which you deigned to read my work and to pronounce your most eagerly awaited judgment on it, as well as especially for the letter, which I think to have received from you will always be a source of the highest glory to me. I am convinced that the praise that you accord me in the very celebrated review, which I had not expected, is due more to your benevolence and singular humanity toward me than to my merits. I will certainly not treat carelessly your advice, especially that concerning the vessel walls; but rather I will now see to it with the greatest care that this matter, whatever it may be, will at last be elucidated by new experiments. I admit that I was too thrifty in reviewing experiments, since for the sake of reducing costs I chose only those experiments from my collection that I thought were exactly sufficient for demonstrating what I wanted. For the rest, I would hardly think that the opinions of Needham could gain strength from my work, illustrious one, since he dealt with another matter entirely different from mine. He strove with his infusoria not to explain but merely to demonstrate the occurrence in nature not of the normal generation of perfect animals, which results from the union of parents, but rather of the birth of microscopic animalcules from putrid matter without the union of similar animalcules. I, however, not differing concerning birth up to this point, attempted not to demonstrate that the usual generation of perfect animals and plants occurs but rather to explain it from its physical causes. But the remaining obscure metaphysical speculations, which Needham adds without any experiments, merit hardly any attention, in my opinion at least. And here too he has nothing in common with me, as, if it were worth the effort, I could easily make amply clear. I explained the fact that a heart does not arise in plants in § 216 indeed with too much conciseness on this basis: in general, the heart arises only if the vessels are so formed that smaller ones are always being gathered into larger ones, and these finally into one common vessel, which, whatever its form may be, is itself the true heart; but in plants, for this reason, collected or branched vessels cannot be formed, and consequently also for this reason a heart cannot result in them. Why branched vessels cannot be formed in plants was explained already in the preceding § 215. I gladly admit that up to this point I have been totally ignorant of the source of irritability; therefore, I would not know how to explain the origin of pulsation either, but I think that the remaining attributes of the arteries have been explained in § 202, 205, etc.

Now, illustrious one, may you live safe and sound for a very long time, for the continuous growth of medicine and of all natural philosophy, both of which sciences have long owed their better part to your labors. How indeed wholly new, more solid, and more pleasing the face of the economy and nature of animals has appeared, on account of your method of investigation! Especially since you discovered the great principle of irritability, which seems to me certainly to be the first foundation of all animality. How I now congratulate the learned world for so great a treasure as it has received in the *Elementa physiologiae*![2] To this the whole of medicine, with whatever form it may ever be enobled or embellished, will always owe the firm, unchanging foundation on which it rests. But I cease saying these things to you, for you are more aware of and see better than anyone else how much natural science owes to you. Now, I instead commend myself to your further favor, illustrious one, for I will always be your servant.

LETTER III

Berlin
20 December 1764

Please do not construe it in bad manner, illustrious one, if I once again dare to disturb you, even if you are busy with weightier matters or perhaps with other, higher investigations of nature. I have undertaken new experiments on incubated eggs so that I might render the theory of generation more firm; and I have paid particular attention to what seemed most doubtful to you. The little book that I am sending you [*Theorie von der Generation,* 1764] will make clear what the particular results of these labors are. Its first part contains those things that it seems to me can be elicited by stricter reasoning from observations already known before, and also from the nature of the thing itself. The second part establishes a briefer theory, which is corrected in places. Finally, I have advanced in an appendix those things resulting from repeated experiments that seemed to me to accomplish the most toward my goal. I have added to these things a drawing of these observations [see Figure 20], which, although not very ornate, is at least true, in order that you, illustrious one, can judge them yourself. Figure 1 shows that very delicate cellular area surrounding the spinal column, in which a short time later those protuberances that are the beginnings of the feet and wings arise out of the new liquid deposited within this area, which is itself the first beginning of the abdomen and thorax. Figure 2 shows the same area, but now it is much more complete, more determinate at its edges, less diffuse, more firm and solid in its substance. From this you will easily perceive in this matter, illustrious one, the mode of origin from the successive secretion of liquid, which formerly appeared lighter and more *diffuse*, but which now is seen to be more solid and contracted. In truth, in this more solid area, those protuberances are now already present that reveal the first beginnings of the extremities. You will see at the same time, illustrious one, to what extent the heart is covered on both sides by the true beginning of the thorax. Its anterior and largest part projects freely, covered by the whole membrane of the amnion, which, bent back and continuous with the skin of the embryo, covered the heart altogether. I have discussed all these matters in the appendix [of my book] on pp. 257–59. Figure 3 is that umbilical area [the area vasculosa], with the first beginnings of vessels, which I have described on pp. 260–61. Certainly these places, which are hardly a little more obscure than the area itself is, and which are full of pale dark fluid, are nothing else but that fluid which clings under the membrane of the yolk, which is precipitated from the white matter that makes up the area and liquefies in various places, and which stagnates; hence the area takes on a white color, opacity, and a character that is quasi-dry. If the liquefying and dissolving of the area continue, the fluid flows together from all directions, and the places where it has been stationary up to now become vessels, or rather streams, as Figure 4 shows. In the interior parts of the area, fissures and ruptures appear in various shapes, at the places where the fluid has not yet flowed together from all sides; in the exterior parts, however, this fluid has already broken up the area to such an extent that it is completely divided into larger and smaller islands, mostly oval in shape. The integrity of the area in the places closest to the

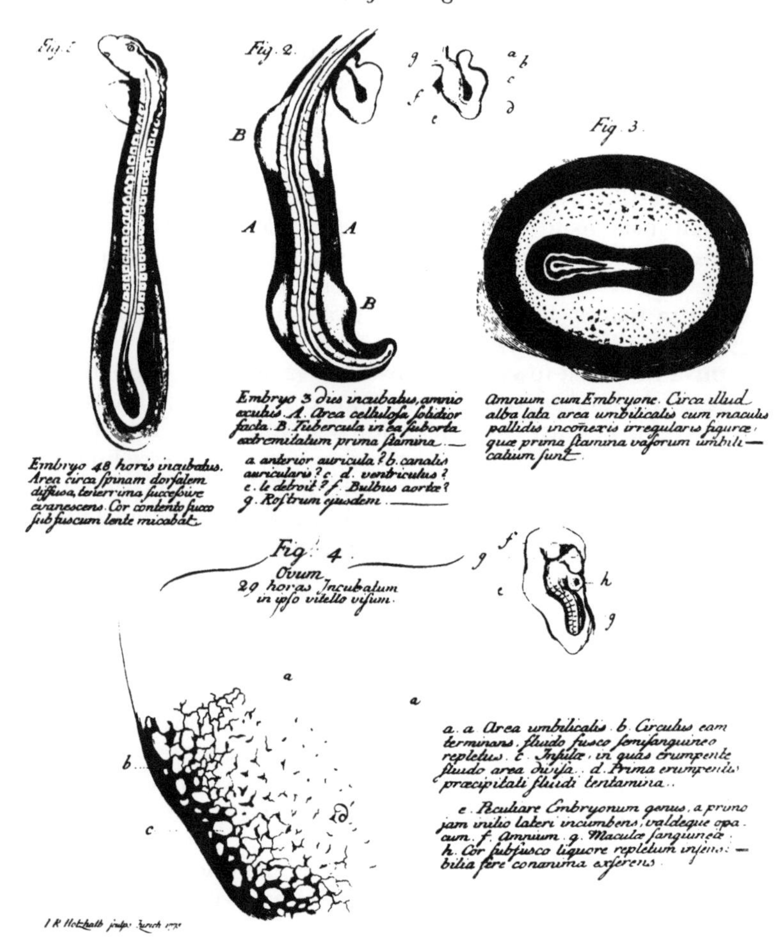

Figure 20. Drawings sent by Wolff to Haller along with Letter III, 20 December 1764. (From Haller, *Epistolarum ab eruditis viris*, 1773–75, vol. 6; courtesy of the Francis A. Countway Library of Medicine, Harvard Medical School. See also Belloni 1971.)

amnion, which cannot be divided even by a needle without evident violent laceration, prevents you, illustrious one, from being able to regard these fissures as perfected vessels that have only been dilated by the force of the heart and by blood propelled from the embryo.

If it should please you, illustrious one, to declare once more your opinion concerning these matters and others that seem more noteworthy to you that are contained in the little book, certainly nothing could happen that would please me more. I will be all the more persuaded of the truth of what you accept, for your most extensive experience and most solid judgment are for me superior to the weightiest arguments. Those things that seem doubtful to you will certainly require further investigation, and will therefore provide opportunity for new experiments and new speculations. Therefore, your advice will always be the most useful of anyone's to me.

As far as the most celebrated Bonnet is concerned, I admit that I perhaps wrote a little too harshly against him; but indeed, in thinking the matter over seriously, I do not seem to myself to have sinned greatly. Certainly it is not the contradictions that affect me. For if he had contradicted me sincerely and openly, or had attacked me in any way, even vehemently or sharply, I would have accepted all this in a rather agreeable spirit, well aware that all vehemence of expression can hardly be avoided in disputes. But Bonnet set a trap, and exhibited signs of an evil mind; I do not think that this matter requires further explanation.[3]

Further, I pray, illustrious one, to the Divine Power that all the days of your life in the future may be happy, and not disturbed by any cares or by the envy of adversaries, so that you may be able happily to finish for your own enjoyment your great work, which the learned world eagerly awaits. I will certainly always rejoice with a most grateful mind that I owe most of my accomplishments to this work and to your other writings. Farewell.

LETTER IV

Berlin
5 May 1765

Although I am always afraid, illustrious one, that by interrupting you when you are occupied with very serious matters I may finally make myself troublesome and hateful, nevertheless the latest letter with which you have seen fit to honor me is of such a sort that, unless I respond to it, I may rightly be accused of impudence or at least of gross negligence. Therefore, I think that I may rightly ask you, illustrious one, to hear my excuses on this occasion at least.

As far as the first circumstance is concerned, about the thoughtless quotation of the review [Haller 1760; reprinted in the *Theorie von der Generation*], I admit that I committed an error that shames me more than I can say. Of all those things that you, illustrious one, have rightly pointed out to me, I was, through some unlucky star, completely unconscious. If any of these things had entered my awareness, I would not have acted so inconsiderately. I certainly intend to be careful in the future not to commit any similar error.

As far as the second matter is concerned, if I had attacked the most celebrated Bonnet when he was unaware of his fault and innocent, I would be very ashamed and would never forgive myself for such a great transgression. But, illustrious one, how can it happen that someone should openly seek to defend one thesis and to refute the contrary position, and that he should deliberate for a long time concerning this matter (as the preface of the book of this celebrated gentleman shows), but that, before he publishes his own book [1762], he should neither read nor have read to him another book, in which the contrary thesis is especially defended, which has only recently been published, in which new arguments are advanced, and which, finally, should certainly have been known to him at least from its review? Everyone reads on such occasions even the most worthless books that discuss the same arguments, so that he may have at least some familiarity with their contents. For these and other reasons I thought that I was obliged to reply to my refuter, who thought me worthy of refutation but not of citation. I should wish that whatever happened in these matters in the meantime had not been done. It is better to bear than to inflict injury, as I now see and have long known; but it is difficult in disputes to avoid all harshness of expression. For when I am eager to explain something, I often neglect the words, and words that seem harmless enough to me as I write them, often seem harsh to a reader, and rightly so. This is the explanation that I am able to give for what has been done up to now and for what cannot be undone. As far as what I should do in this matter in the future, that depends upon your will, illustrious one, and upon that of the most celebrated Bonnet.

I saw the beginnings of the vessels in the umbilical area [area vasculosa] shortly after the first day of incubation and for the whole second day, when the area, examined with the naked eye, appears wholly intact and equal, and the vessels appear to have no color, not even a dark one. Then with the microscope I discovered spots that were somewhat darker than the rest of the area, and irregular in shape and size and not connected, which subsequent observations revealed were the first delineations of the vessels. When in fact on the third and following days red or dark points and lines begin to be apparent to the eyes, then the microscope easily reveals connected and branching roads that already have a greater similarity to vessels.

Spirit of wine also binds the two layers of the umbilical membrane together in such a way that the places that contain dissolved fluid and are the beginnings of the vessels project out on the lower surface; for this reason, the filled vessels swell and jut out beyond the parenchyma. In this way a condition that is a little more perfect than it is in truth is produced in the area. For after the first day, when the spots are still not connected, wrinkles are for this reason formed on the lower surface that are varied and mostly parallel, which previously did not exist; but on the following days what had earlier been branching roads appear, under the effect of the binding spirit [of wine], to be swollen, branching vessels. But even at this time, these swollen vessels seem to me to be nothing other than the lower layer of the area, distended by fluid in these places and raised higher, since the other places, where the white matter adheres between the two layers, producing the interstices between the roads, are more constricted and therefore lower.

I, like all who study nature, await with great longing the expanded Latin edition of your most excellent work on the formation of the chicken and of bones [in *Opera minora*, vol. 2]. That day will be an especially festive one for me on which I begin to read and understand, in the long-desired volume VIII of your *Elementa physiologiae*, the universal history of generation, which is a matter so close to my heart.

Farewell, illustrious one, and, if there is one thing I beg of you, never deprive me of the favor of which you have considered me worthy up to now.

LETTER V

Berlin
16 November 1765

On the 13th of November I received the letter that you wrote on the 27th of September, illustrious one, and I perceived in it new indications of your customary humanity and benevolence toward me. If by chance I have erred in those things that were concluded by reasoning, I do not think that it is a dishonor to me, so long as my mind has always been hostile to hypotheses and eager to pursue the truth itself. But I seek the praise that I know how to venerate justly your most solid doctrine and you, its illustrious author, and how to form an accurate judgment of and to acknowledge gratefully all the things – how great they are! – that I owe to you, illustrious one, in the physical sciences.

I understand that the formation of vessels in the egg, as well as the whole matter of generation, is a thing of great importance and certainly deserving of further investigation. Therefore I shall first await the new edition of your work [*Opera minora*] and read of the new experiments, and I shall then, following your footsteps, see what my shoulders can bear in further investigating nature itself.

Farewell, illustrious one, and continue, as you are accustomed to do, to honor me with your favor.

LETTER VI

Berlin
6 October 1766

After reading the last volume of your *Elementa physiologiae* [vol. 8], illustrious and most excellent one, I considered waiting for the second volume of your *Opera minora*, so that, having obtained full knowledge of your arguments, I could either concede whatever controversies exist between us, or explain what still remained dubious! but, since I see that the second volume will not appear soon, and since I am unable to repress the feelings of my most grateful mind any longer, I have therefore because of this decided rather to write to you now.

First of all, illustrious and most excellent one, I thank you as much as I can because you granted a place for my thoughts, such as they are, concerning generation in your immortal work, and because you praised my small service, in investigating this difficult matter of nature, and my efforts.

I thank you further because I see you are benevolent toward me and because indeed I see that you love me, greatest one, although you have never seen me and have known me and the condition of my mind only from letters; and indeed this you do only for the sake of truth and virtue. May God repay you for this; for I do not hope that I will ever be so important in this life that I will be able to do anything for you that is worthy of your virtue, unless you are willing to consider the eternal veneration of a mind for you.

As far as our controversy is concerned, I think this way. The truth is no dearer to me, illustrious one, than to you. Whether organic bodies of nature are evolved from an invisible state to a visible one, or are truly produced, there is no reason why I should choose this one rather than that or why I should prefer that one, not this. And the same is also your view, illustrious one. The truth alone we both pursue; we search for what is true. Why then should I dispute against you? Why should I resist you, when you strive with me toward the same goal? I would rather entrust epigenesis to your care, to defend and perfect, if it were true; but if false, it would be an odious monster to me also. I would admire evolution [preformation], if it were true, and I would worship most submissively the revered Author of nature, the Divine Power which is inexplicable to the human intellect: but if false, you also would quickly reject it, even if I were silent.

While I attributed a double layer to the membrane of the yolk, concerning which you write, illustrious one, in the *Elementa physiologiae*, p. 95, I have recognized that in a yolk incubated for 14 and 16 days, as you know, the exterior layer of the yolk sac, smooth like the peritoneum, is continuous with the mesentery, whereas the interior layer, which is distinct, soft, wrinkled, like the villous membrane of the intestines, is, analogously, continuous through the yolk duct with the villous membrane of the intestines. But the unincubated yolk appears to me to enjoy just one membrane, not two, unless one considers the membrane that belongs not to the yolk but is only the interior leaf of the umbilical membrane [allantois], as you have written, illustrious one, in the place cited above. This is certainly after 10 or 12 days the umbilical membrane, as you have called it, or the chorion of Malpighi, and it is not continuous with the intestines; but in an unincubated yolk, it seems to me that its covering is in fact borrowed, or rather that it takes the place of it [the yolk membranes later seen to be continuous with the intestines], since this single membrane contains the fluid substance of the yolk. I have seen when I immersed a piece of this [membrane] that I had cut off in water or in spirit of wine, that lying under it was a pulpous, soft substance, from which the thin [membrane] then spontaneously separated, but this appears to me to be a part of the absorbed albumen thickened by the water or spirit of wine.

Do not Fabricius ab Aquapendente and Harvey seem to have the same opinion, since the latter discourses in Exercise 12, on the remaining parts of the egg, on the middle of p. 42 [of Harvey 1651] concerning the yolk as follows: "It is the softest liquid of the egg, covered by a *most delicate* membrane, which flows away if that membrane has been torn." For I believe that if they had any opinion other than this common one, they would have described it more accurately.

I have looked for the vessels of yolks in the ovary of a pregnant hen, and I have found that they derive from the vessels of the ovary, that they pass over to the yolk with the stem of the yolk, that they disperse over its surface, almost as the ciliary arterioles next to the optic nerve pass over to the bulb of the eye and disperse there, except that the shoots are much shorter and more numerous.

But these things, such as they are, since they establish nothing certain concerning generation, let us omit entirely. I await the second volume of your *Opera minora*; then I will begin to inquire into other things.

In the meantime, I beg most earnestly only one thing from you, illustrious and most excellent one, that you may continue to honor me with your favor, as you have honored me up to now.

Farewell.

LETTER VII

Berlin
17 April 1767

A severe inflammation of the eyes from which I suffered prevented me from immediately answering your letter, which was full of your accustomed favor toward me, illustrious and most excellent one. Now I admit that the arguments that you have presented for elaborating the hypothesis of evolution, are altogether of the greatest weight, and seem to me to be so powerful, that I have almost no idea what I shall do in the future with regard to improving the theory of generation. Although it has not been hidden from me that the evolution of natural organic bodies can be brought forth against opponents as outstanding evidence for a Divine Author, which would be destroyed if epigenesis were demonstrated, nevertheless I had not perceived the whole matter, as you expounded it to me, highest and most worthy one, and I had not considered all things sufficiently. For I now see that the question concerns less the demonstration of the truth of religion and more how this demonstration can be made simple, short, and evident, and so constructed that no deceit can easily be built by plotters against it. I do not think that any other argument is so apt for this purpose as evolution [preformation], nor anything more opposed to it as its refutation. In fact it is true that nothing is demonstrated against the existence of Divine Power, even if bodies are produced by natural forces and natural causes; for these very forces and causes and nature itself claim an author for themselves, just as much as organic bodies do. But by far the clearer and better the proof would be, if, in contemplating the state of nature, we were to find that a single product of it or organic bodies had had need of the Creator, and that nothing organic could have been produced through natural causes. But perhaps I will sometime see all these things explained in the second volume of your works [*Opera minora*].

Now, illustrious and most excellent one, I have been called to St. Petersburg to the position of Professor of Anatomy and Physiology and of member of the Academy of Sciences with a salary of 800 Rubels. I did not long consider what I should do; I accepted. I have received 200 Rubels to cover the costs of the journey, and I will set off from here in a few days. I therefore earnestly beseech you, illustrious and most excellent one, to continue to think me worthy and to commend me in the future, even though I am quite far away from you, with the same honor and favor that you have found me worthy of up to now. For nothing can be more to the honor and glory of a physician, or of anyone at all, and therefore of me also, I have always believed, than to have received letters from you, highest one, or to be mentioned by you in your immortal writings, or to be favorably known to you. I therefore greatly desire and earnestly beg that I should receive from you in the future that very thing that I have received from you up to now as the most fervently wished-for reward of my labor and attempts.

I am sorry that I could not receive from you yourself the second volume of your *Opera minora*, as you generously promised, since the shortness of time stands in the way, and no merchant at Göttingen or Leipzig is known to me. In the meantime, I give greatest thanks to you for your good will, and I will therefore wait until the book is published.

Fare well in the meantime, illustrious and most excellent one, and may you long, very long, enjoy a tranquil life.

LETTER VIII

St. Petersburg
27 September/ 8 October 1776

Since I happened to make some observations concerning the foramen ovale that I judged were not altogether unworthy of your attention, I am sending to you, illustrious one, this latest volume of our *Novi Commentarii* [1775], which is designated as a commemoration of the fiftieth anniversary of our Academy, which the *Acta Academiae* will now follow in the future, and of which my dissertation on the aforementioned foramen forms a part.[4]

This foramen[5] has been placed (as you well know) by the most celebrated anatomists between the two sinuses[6] of the heart and has been so thought to belong to the right no less than to the left that they have said that its valve is composed of a double layer, of which one layer is continuous with the internal membrane of the left sinus and the other with the internal membrane of the right sinus. This foramen, to express myself in one word, I have found to be nothing else but the left orifice of the inferior vena cava, whose superior lumen is divided by the well-known arch into two orifices. By the right one of these the way is opened into the right sinus, and by the left one into the left sinus. Here the valve has one layer that is continuous with the left sinus and another layer that is continuous with the inferior vena cava.

There is no need to tell you, illustrious one, since you see all this at first glance, how much this structure, even though it has been expressed in few words, differs from the common notion of the foramen ovale, nor how much different it makes the movement of blood in the fetus, as has been briefly explained in this review of the dissertation and at greater length in the dissertation itself.

I think that this fabric escaped even the most acute anatomists up to now because of the difficult and complicated structure of the organ, because of the long-standing idea of the foramen ovale, and especially because of the usual method of dissecting. You will recognize, illustrious one, that it is as I have said as soon as you open the right sinus in a fetus together with the right ventricle of the heart, if the inferior vena cava is left unharmed.

I do not think that you were particularly involved in those disputes that once took place in the Paris Academy of Sciences, except to the extent that your great name reigns throughout all anatomy and physiology.[7] How I should rejoice if finally in some particular matter, honored with your judgment, great one, I should be permitted to agree with you! I earnestly desire to agree with you not only in the theory of generation but also in every truth of even the smallest weight.

For it is proper that I be held by these wishes, since there is only one truth in each matter.

I honor the illustrious Bonnet, as I always truly have, with the highest respect. I rejoice to be about to have the occasion to attest this publicly.[8] The very rich storehouse of monsters that has been collected and preserved over a long series of years in the Imperial museum has now been handed over to me, so that I can compose a description of them and perform anatomies where I decide to. In this therefore it will be necessary to deal once more both with the origin of monsters as well as with generation in general.

I have not written to you for many years, illustrious and excellent one. This has been because of the long journey that letters would have to make through many hands and especially because in all this time, since I have often been prevented by sickness, I have written nothing that could have served as a sufficiently worthy approach to you.

Farewell, and, as you have done up to now, highest one, continue to honor me with your favor.

LETTER IX

St. Petersburg
7 May 1777

I received your letter, which is dearer to me than anything else, on the 10th of April, illustrious one. Together with the whole learned world and with all good men, I grieve, illustrious one, over your bad health! But I hope that without doubt God, who is so great and good, will soon restore your complete health.

On the 11th of April our Academy also received a response from you. I give you indebted thanks, illustrious one, because you were willing to add your most illustrious name to the splendor of our Academy, a signal honor in which I too participate.

My observations center chiefly around the mature or recently born fetus of humans and of calves; in these I have found that the foramen ovale, which was thought to be in the septum of the sinuses, belongs to the inferior vena cava, which ends on top with two orifices or short branches, and that with the right one of these it enters the right sinus, whereas with the left one, which is itself the foramen ovale, it enters the left sinus. And yet, in my opinion at least, illustrious one, all this is most consistent with your most beautiful observations on the incubated egg, which made it clear that only one auricle is present in earliest life together with the left ventricle. Without doubt, this is the left auricle, since it empties into the left ventricle. It is therefore necessary that the inferior vena cava should extend into this auricle in this first stage, since the right auricle is not yet present. And it is certainly this structure whose vestiges I recognized when I observed a fetus of nine months. In it, the inferior vena cava still crosses into the left sinus through the left orifice with half of itself, and with half it crosses into the right sinus; whereas later, after the left orifice has been closed, it is inserted in the right sinus alone. Indeed, closer to your observations, illustrious one, is the structure that I saw in a human fetus of three months, where the inferior vena cava passes directly into the left sinus by a large orifice (which is the same foramen ovale), which is not opened by a valve but which does have a small lateral orifice, wider and stopped by the Eustachian valve, by which it opens into the right sinus, which at this stage is already very large.

I know that these things cannot be sufficiently reconciled with the things that anyone who practices customary dissection has observed in nature, or thinks he has observed. But the customary dissection, which we practice most frequently, and which is most suitable and indeed necessary for uncovering and exhibiting the foramen ovale, does not appear to be sufficient for recognizing its position more deeply and for explaining the phenomena. In most illustrations concerning the foramen ovale, as in Table 1, Figures 8 and 20 in Trew [1736] and in that very beautiful illustration that is Figure 1 in the Table concerning the foramen ovale in Fascicle 4 of your immortal *Icones anatomicae* [1743–54], illustrious one, the right sinus is exposed from the posterior, or right, side *along with the inferior vena cava.* In these illustrations then appears immediately the orifice that is commonly called the foramen ovale. All this region in which this orifice is found is believed to belong to the right sinus, and that part of it that is visible is believed to be that which produces the septum between the two sinuses. Therefore, the foramen ovale is thought to be in the septum of the sinuses and to be observed and seen in the septum of the sinuses.

Now I ask, illustrious one, where in these illustrations or in nature, which they represent, is the boundary between the right sinus and the inferior vena cava that allows us to determine that the foramen ovale belongs to the sinus and not to the vein? And where in these illustrations is the orifice by which the inferior vena cava opens into the right sinus? by which alone this vein can be distinguished from the sinus, which cannot appear in these illustrations, since the anatomist destroyed the orifice by cutting into the vein? So that it can be determined whether the foramen ovale is in the sinus or in the vein, the sinus must be cut from the anterior side together with the anterior ventricle and the venous orifice of the ventricle (as in my Figure 1 [see Figure 21]), *leaving untouched the inferior vena cava and the orifice* by which it opens into the right sinus. Then in the open right sinus will first appear this orifice of the vena cava, by which it opens into the right sinus (my Figure 1, numbers 10, 11, 12), above the isthmus of Vieussens (10) and below the movable edge of the Eustachian valve (this valve is separated from the isthmus in the aforementioned illustrations and bent back outside its position; in this way, the orifice of the vena cava disappears). Within this orifice and in the very cavity of the inferior vena cava, in its left wall, the foramen ovale, or rather only its larger, higher part, appears (my Figure 1, nos. 13, 14, 15, and Figure 5, x, y, z), since the lower part of the ring and foramen, descending deeper in

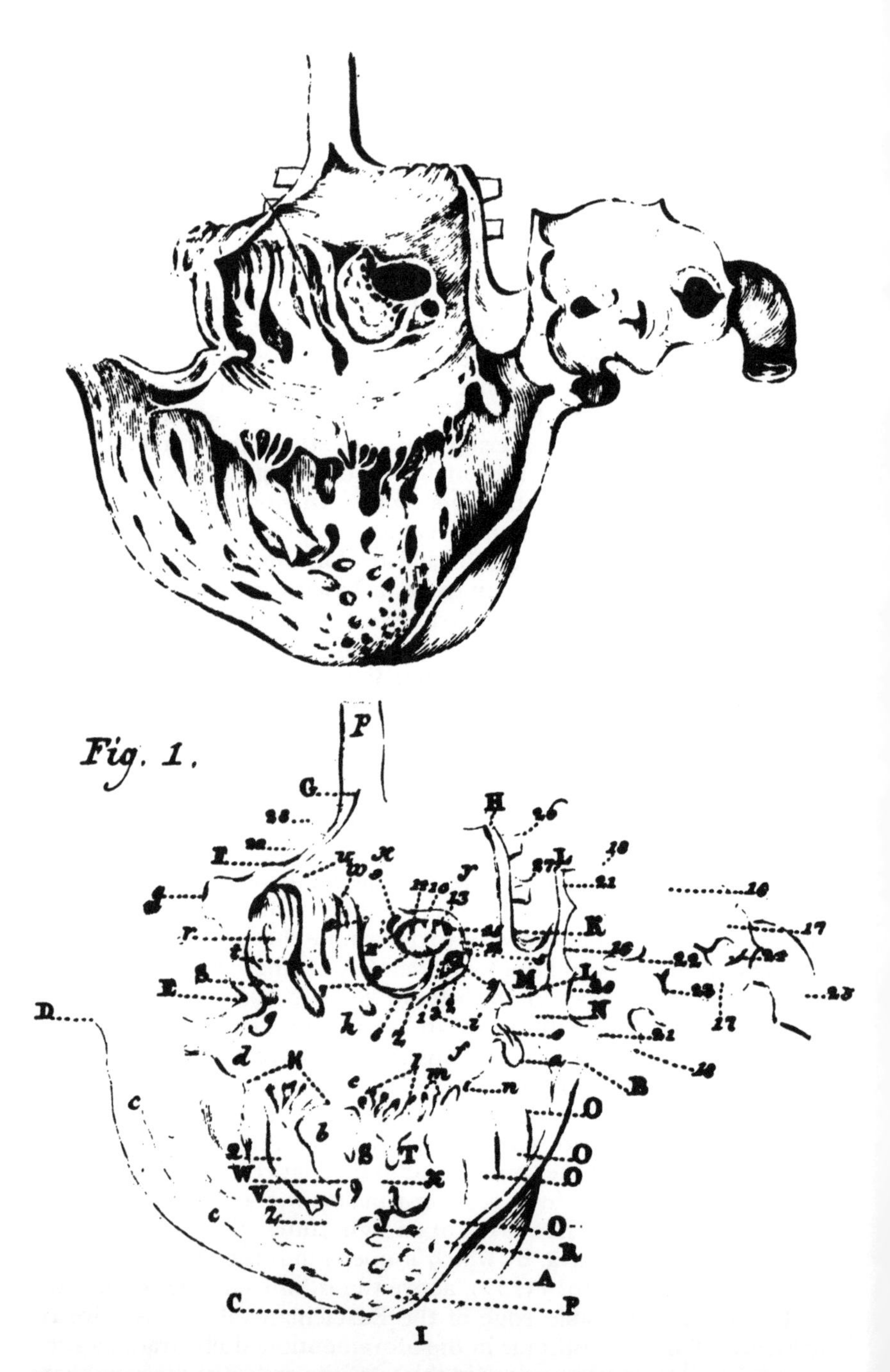

Figure 21. One of Wolff's illustrations from his paper on the foramen ovale, showing the right auricle and ventricle of the heart, opened from the front right side. *B, C, D,* right ventricle; *E, F, G, H, K,* right sinus (auricle); *I,* apex of the heart; *p,* superior vena

the wall of the vein, is covered by the Eustachian valve, and the whole ring and foramen ovale only appear when the vena cava is cut, as in your splendid Figure 1 and also in your incomparable eighth Figure of the same Table, in which you exhibited for view the Eustachian valve, illustrious one, loose and bent back, and the fossa ovalis which was filled in in this way, as was proper. Therefore the foramen ovale belongs to the inferior vena cava, in whose wall it is located, and not to the septum of the sinuses, which is bounded by the highest part of the ring, or isthmus (my Figure 1, no. 10). It is nothing else but the orifice of the inferior vena cava, by which it opens into the left sinus.

The following, illustrious one, are therefore the primary arguments of my dissertation, which I do not think were ever stated before: 1) The inferior vena cava terminates on top in a fetus of nine months in not one but two orifices or short branches. 2) By one of these orifices it enters the right sinus, by the other, the left. 3) The so-called foramen ovale is nothing other than this very left orifice of the inferior vena cava, by which the way is opened into the left sinus. 4) The sinuses therefore do not communicate with each other, but each of them communicates separately with the inferior vena cava. 5) And the blood therefore does not pass from the right sinus into the left, nor from the left into the right, but it proceeds directly from the inferior vena cava partly into the right sinus and its auricle through the right orifice, and partly into the left sinus through the left orifice or foramen ovale. From the right sinus and auricle, in fact, *all* blood passes into the right ventricle, as it passes from the left sinus into the left ventricle.[9] I think that the rest, if anything remains, is of less importance. Please excuse, illustrious one, the loquacity with which I have abused your patience.

If you wish to honor me with the second part of your *Bibliotheca anatomicae* [1774–77], illustrious one, please send it to Basel to Emanuel Turneiser the bookseller, who will take care to have it sent on to St. Petersburg.

Farewell, illustrious one! I pray to immortal God that you may be strong! and that you may still live safe and sound for a long series of years and may teach the learned world.

Figure 21 (*cont.*).
cava; *1, 2, 3, 4, 5*, orifice of the coronary vein; *5, 7, 8, 9*, Eustachian valve; *8, 10, 11, 12*, right orifice of the inferior vena cava; *13, 14, 15*, cavity of the inferior vena cava, in which a part of the left orifice (or foramen ovale) can be seen. (From "De foramine ovali," 1775)

Notes

Chapter 1. Introduction: mechanism and embryology

1 As both Roger (1963:325–26) and Bowler (1971:221–22) have pointed out, some confusion has existed over the relationship of the terms "preformation" and "preexistence." In particular, they argue that using "preformation" to denote 1) early seventeenth-century theories that held that the embryo is preformed in the parent before conception *and* 2) eighteenth-century beliefs that all embryos have existed from the Creation, has led to a mistaken belief that the later theories grew directly out of the former. Thus, Roger and Bowler suggest reserving "preformation" for the earlier school of thought and "preexistence" for eighteenth-century theories based on preexisting germs. As I shall be dealing only with preexistence theories in this book, however, I shall use these two terms interchangeably, meaning in no way to detract from the valuable clarification that Roger and Bowler have brought to the history of embryology during this period. (For a dissenting voice with regard to the terminology suggested by Roger and Bowler, see Wilkie 1967:147–50.)

2 Both Aristotle and Galen believed that the embryo emerges gradually from the developmental process. Aristotle maintained that the female contributes the material from which the embryo develops (and thus the material cause), while the male semen, acting as the efficient cause of reproduction, provides only the form (and thus the formal and final causes). Galen subscribed to a version of the Hippocratic two-semen theory, in which the embryo develops out of material derived from the bodies of both parents. For more on the embryological views of Aristotle and Galen, see Preus 1970, 1977.

3 On Harvey's theory of generation, see also Webster 1966–67; Bodemer 1968; and Foote 1969.

4 See the French–English edition of Descartes's *Treatise of Man* (1972), with Hall's invaluable editorial comments. See also Hall 1970 and Sloan 1977.

5 See Roger 1963:335; Adelmann 1966, 2:885–86, 915–20; Bowler 1971. Haller, for example, remarked in 1744, "But the theory of evolution [preformation] proposed by Swammerdam and Malpighi prevails almost everywhere" (1739–44, 5, pt. 2[1744]:498, n. *g*; trans. Adelmann 1966, 2:893). (See Chapter 2, note 5 on the use of the term "evolution" for "preformation.")

With regard to Malpighi's observations, it is likely that the summer heat in Italy prompted the initial development before incubation of the eggs he was observing, in addition to the development that normally takes place before the egg is laid by the hen. Malpighi's descriptions of the stages of chick development are also frequently ahead by a few hours, a fact that was noted in the eighteenth century and frequently ascribed to excessive summer temperatures.

6 See Adelmann 1966, 2:819–1013, for a discussion and translation of Malpighi's two treatises on embryological development.
7 See, for example, Roger 1963:325–26, 343–44; Oppenheimer 1967:129–32; Gasking 1967:42–50; and Bowler 1971:235–36.
8 De Graaf (1672) dissected the ovaries of numerous mammals and observed the morphological changes accompanying ovulation. He correctly described the expulsion of the egg into the Fallopian tube, but he thought that the entire ovarian follicle (now called the "Graafian follicle") constituted the egg, having not observed its rupture to release the tiny ovum. Karl Ernst von Baer was the first to see the mammalian ovum and to correctly describe its production (see von Baer 1827, 1956 trans.).
9 It is because of Buffon's denial of actual development after the initial formation of the embryo that some commentators have pointed out that his theory is not a truly epigenetic one. See Cole 1930:103–4; Roger 1963:546, 548, 556–57; Castellani 1972; and Bowler 1973.
10 See Glass 1959b:52; 1974:186; Roger 1963:458, 469, 528, 556; 1973:577.
11 See also Castellani 1969-70; Pancaldi 1972; and Farley 1977:24–27.

Chapter 2. Haller's changing views on embryology

1 For biographical information on Haller, see especially Zimmermann 1755; Condorcet 1780; Beer 1947; and Hintzsche 1959, 1972.
2 Original German text: "Ja, in dem Samen schon, eh er das Leben haucht/ Sind Gänge schon gehölt, die erst das Thier gebraucht."
3 The *Praelectiones academicae* (Haller 1739–44) was based on Boerhaave's popular textbook *Institutiones medicae* (1708 and many later editions). Haller's edition was designed to elaborate on Boerhaave's tersely worded doctrines, and he included two sets of notes. The first set (which forms the bulk of the text) explains and amplifies Boerhaave's views. The second set, which appears as footnotes to the first, contains both additional information and critical comments by Haller.
4 It is estimated that Haller published ten- to twelve-thousand anonymous reviews, mostly in the *Göttingische Anzeigen von gelehrten Sachen* and, to a lesser extent, the *Bibliothèque raisonnée*. Pioneering work on identifying and classifying Haller's reviews has been done by Karl S. Guthke (1926b: 37–49; 1970; 1973; 1975:333–53; the latter two publications include lists of Haller's reviews in the *Bibliothèque raisonnée*). The two most secure methods of establishing Haller's authorship of any given review are its having been reprinted in Haller's *Sammlung kleiner Hallerischer Schriften* (1756 and 1772b eds.) and, for the *Göttingische Anzeigen von gelehrten Sachen*, the existence of an "H" in the margin of Haller's copy of this journal, now in the Stadtbibliothek in Bern. (Concerning Haller's reviews after 1769, see also Fambach 1976.) For reviews discussed in this book, the method I have used in determining Haller's authorship is indicated following each review's listing in the Bibliography.
5 In the eighteenth century, the term "evolution" was often used, by Haller and others, to denote development from preexisting parts. At this time, this use had no connection with the evolution of species (see Bowler 1975; Gould 1977:28–32). According to Cole (1930:86), Haller was the first person to use the term "evolution" as the equivalent of preformation (see quotation in Chapter 1, note 5).
6 The editions of the *Primae lineae physiologiae* provide a guide to Haller's changing views on embryological development. The first and second

editions (1747, 1751c; trans. 1754) support epigenesis, while the third (1765b; trans. 1803, 1966) and later editions support preformation.

7 See, for example, Jenny 1902; Cole 1930; Rostand 1930; Bilikiewicz 1932; Needham 1934, rev. ed. 1959; Meyer 1939; Schopfer 1945; Adelmann 1966, 3:1388; and Hintzsche 1972. For a review of the literature on Haller's conversion from epigenesis to preformation, see Mazzolini 1977.

8 This footnote does not appear in the first edition of the *Sammlung* (1756).

9 See also Gasking 1967:107–16; Toellner 1971:182–88; and Roe 1975.

10 Regarding Haller's authorship of this review, see note 4.

11 It is interesting to note that Maupertuis, whose theory of generation was very similar to Buffon's in this regard, recommended performing a series of experiments where a specific part of an organism's body would be destroyed generation after generation to see if deformed offspring would result. There is no evidence that these were ever carried out (see Maupertuis 1745, 1966 trans.:78–79). In the late nineteenth century, August Weismann tried a similar experiment to test the inheritance of acquired characteristics. He cut off the tails of mice for several generations, but without any visible hereditary effects (see Weismann 1891–92; Churchill 1976).

12 The "tree of Diane" was frequently cited in the eighteenth century as an example of how natural laws could produce complex, organized formations. A treelike figure forming on the surface of water through the chemical interaction of silver, nitric acid, and mercury, the *arbre de Diane* was used by Maupertuis (see Chapter 1) and others as a model for epigenetic development.

13 This passage is followed by a statement that has led Roger (1963:707) and Toellner (1971:184–88) to erroneously date Haller's conversion to preformation as having taken place by 1751. Here Haller states, "This constancy [of same type of organism resulting from reproduction] has convinced me against all the experiments of M. Needham [on the formation of animalcules in infusoria]; there must accordingly be something prepared and built in the fertile liquid of man and animals, although it is not a miniature of a whole body nor a caterpillar of an imagined butterfly" (1752a:xvi). The "imagined butterfly" is undoubtedly a reference to Swammerdam's demonstration in 1669 that a butterfly could be found folded up in both the caterpillar and the chrysalis, which was taken as a proof of preformation (see Chapter 1). Read in context, this passage from Haller's preface indicates only that Haller believed that the material from which the embryo develops, which is derived from the parent, influences the type of organism that results and ensures constancy of species. That Haller had not yet converted to preformation is further borne out by the fact that Haller reiterated his support for epigenesis in a letter he wrote to Bonnet in 1754 (discussed later in this chapter).

14 See also Haller's discussion of Spallanzani's work in the Addenda to volume 8 (1766) of his *Elementa physiologiae* (pt. 2, pp. 216–17). For two letters from Needham to Haller, in which Needham defends his epigenetic views against the charge of materialism, see Mazzolini 1976. Haller's objections to Needham's theory are also discussed by Duchesneau (1979:80–85).

15 See Haller 1765b, §§ 400, 402, 404, 407. In the *Elementa physiologiae* (1757–66, 4:446), Haller referred to irritability as the *vis contractilis musculis insita* ("contractile force innate to muscles").

16 See Rudolph 1964; Toellner 1967; and Roe 1981.

17 Kuhlemann's dissertation (1753) supported a mild preformationist position, based on a coagulation model; yet because it was not written until after Haller left Göttingen, one cannot conclude from Kuhlemann's views what Haller's own thoughts on generation were at this time.

18 Haller's letters to Bonnet are preserved in the Bibliothèque Publique et Universitaire, Geneva, while Bonnet's letters to Haller are housed in the Burgerbibliothek, Bern. References to this correspondence will indicate "Haller MSS" or "Bonnet MSS," followed by the date of the letter. (See also Bibliography, "Unpublished Materials," for the titles of these collections.) For information on Bonnet, see Whitmann 1895a, 1895b; G. Bonnet 1929; Savioz 1948a, 1948b; Castellani 1972; Bowler 1973; and Anderson 1976.

19 I am indebted to Renato Mazzolini for correcting a translation I previously suggested for this passage. Mazzolini (1977:226), however, argues on the basis of this statement that, at the commencement of his observations on incubated chicken eggs, Haller had placed himself in a position of "voluntary neutrality" with regard to preformation versus epigenesis. Yet, although Haller clearly recognized that his objections to Buffon's theory challenged his own epigenetic account of development, he still believed in 1754, as this passage demonstrates, in an epigenetic coagulation model for embryological development. (For a similar interpretation of this passage see Bonnet 1762, 1:139; Schopfer 1945:83–84.) This model was transformed into a preformationist one only subsequently, culminating in Haller's conversion to preformationism in 1757.

20 It is not entirely clear what opinion of Bonnet's Haller is referring to here. In the correspondence now extant in Geneva there is no letter from Bonnet to Haller between Haller's letters of 26 November 1754 and 4 January 1755. Bonnet, who reprinted this quotation in his *Considérations sur les corps organisés* (1762), preceded it with the following: "I had always thought that a glue that appeared to organize itself was already organized. I had never been able to accept in my mind [the idea] that the parts of a plant or an animal should have formed themselves successively. The more I reflected on such a formation, the more I felt the insufficiency of the mechanical means celebrated with such complacency by diverse authors. I insisted on the above to M. de Haller, when I received this response" (1:139–40).

21 For additional brief discussions of Haller's observations on incubated chicken eggs in his letters to Somis, see Hintzsche 1965:17, 27, 29. References to these observations can also be found in Haller's letters to Gessner (Sigerist 1923:239, 250, 256, 257, 258–59, 265, 267, 269), Morgagni (Hintzsche 1964:65, 72, 76, 81, 84–85), Caldani (Hintzsche 1966: 18–19, 21–22, 27, 28, 29), and Tissot (Hintzsche 1977:42, 43, 59).

22 Haller believed that conception occurs when an "odorous"portion of the seminal liquor stimulates the preformed heart to begin beating. Although he was well aware that Harvey's experiments on deer had shown that no male semen could be found in the uterus after copulation, a conclusion confirmed at least in part by Haller's own observations on sheep with Kuhlemann, Haller was not convinced that the semen never reaches the uterus. Yet he thought that only a vaporous portion of the seminal liquor actually acts to stimulate the heart's irritability (see Haller 1757–66, 8[1766]:18–23, 154–55). As Mazzolini (1977:203, n. 54) has noted, Haller's theory of conception cannot properly be described as one of "fertilization" but is rather one of "stimulation" or "irritation." Although I have

continued to use the terms "fertilized" and "unfertilized" when discussing Haller's theory, they should be understood in the broader sense of "fecundated" and "unfecundated," the terms actually in use in the eighteenth century.

23 Both Gasking (1967:112–13) and Duchesneau (1979) point to Haller's views on the interdependence of organ systems and on the relationship between organic function and structure as having played influential roles in his adoption and elaboration of a preformationist theory. See especially Duchesneau (pp. 90–100) for an analysis of Haller's explanation of the developmental process in terms of the structure–function relationship.

24 Gasking (1967:110–11) and Mazzolini (1977:199) assert that Haller's membrane-continuity argument was not a proof of preformation per se but only of *ovist* preformation (as opposed to animalculist preformation). Consequently, they argue, Haller's discovery of the membrane-continuity proof was not, as Cole (1930) and others have argued, the cause of his conversion to preformation. Although I agree that Haller's formulation of this proof was only one of many factors contributing to his conversion (as I have shown in this chapter), nevertheless Haller's proof, by demonstrating that the embryo exists before conception in *one* of the parents independently of the other, was for that reason an argument for pre-existence. Its discovery was only one of the observational factors that played a role in Haller's new explanation of development, yet its importance should not be underestimated. For further analysis of Haller's membrane-continuity proof, see Chapter 3.

25 Mazzolini (1977:228–29) suggests that Haller's reading of Diderot's *Pensées sur l'interpretation de la nature* (1754) was influential in his conversion to preformation. Mazzolini reports that Haller read this work in 1757 and that he remarked about it in a letter to Bonnet, "they want to tear from us the strongest and most popular demonstration that we have of the existence of God" (Haller MSS, 6 July 1757). In this work, Diderot had briefly presented Maupertuis's theory of generation (as explained in the *Système de la nature* [1751]), had criticized its materialistic implications, and had hinted at his own theory of sensibility, which would be expanded in *Le Rêve de d'Alembert* (1769) into Diderot's materialist view of vital functioning and of generation. There is no doubt that Haller was disturbed by these views and that, in 1757, they reaffirmed the dissatisfaction with epigenetic theories that had stemmed from his critique of Buffon's and Needham's views in 1751. Whether Haller read Diderot's book before or after he had actually formulated his preformation theory is at this point not possible to conclude for certain.

Chapter 3. The embryological debate

1 Wolff's letters to Haller (nine in all) are reprinted in English translation in Appendix B of this book. All references in the text will give page citations to the original Latin of these letters as published by Haller (1773–75). Unfortunately, Haller's letters to Wolff are no longer in existence.

2 For biographical information on Wolff, see Kirchoff 1868; Wheeler 1899; Schuster 1936; Schütz 1947; Uschmann 1955; Herrlinger 1959, 1966; Gaissinovitch 1961, 1978; Rajkov 1964; and Lukina 1975. For a contemporary view of Wolff, written by one of his students, see Mursinna's letter in Goethe 1817–24, 1:252–56.

3 Wolff's vesicle theory cannot be considered a precursor of the modern cell theory. Rather, Wolff's belief that all organic parts acquire vesicles and vessels during development is entirely consistent with the widespread belief in the eighteenth century that "cellular tissue" (i.e., connective tissue) is the major structural component of organisms (see Wilson 1944). Wolff stressed that the same process of vesicle and vessel formation occurs in plants and in animals, and that one can therefore consider as analogous plant "cells" (observed by Malpighi and Grew) and animal "cellular tissue."

4 With his theory of plant development from a "vegetation point," Wolff anticipated Goethe's concept of metamorphosis. Goethe was apparently unaware of Wolff's ideas until after he had formulated his own theory (see Goethe 1817–24, 1:80–89; see also Kirchoff 1867; Kohlbrugge 1913; and Arber 1946).

5 See Adelmann 1966, 3:1106–27, for these early observers' comments on the area vasculosa.

6 These islands must not be confused with the blood islands (first correctly identified in the nineteenth century) that form in early stages of the area vasculosa. Wolff's islands represent the material broken up by the formation of blood vessel channels, these channels forming *between* the islands.

7 Haller probably received Wolff's dissertation in February, for in March he remarked in a letter to Bonnet, "A young physician of Berlin has sent me a thick thesis in which he defends epigenesis. . . . his name is Wolff" (Bonnet MSS, 4 March 1760).

8 Regarding Haller's authorship of this review, see Chapter 2, note 4.

9 Haller had continued his observations on incubated chicken eggs and on quadrupeds in the summer months of 1763 and 1764, intending to include these in a revised Latin version of his *Sur la formation du coeur dans le poulet,* scheduled to appear in volume 2 of his *Opera minora* (1762–68). The series of observations undertaken in 1765 was designed to refute Wolff's claims as presented in the *Theorie von der Generation.* See Haller's unpublished journal of experiments, "Observationes anatomicae Bernenses," vol. 3.

10 Adelmann (1966, 3:1028–29) identifies Maître-Jan's "little white body" as the nucleus of Pander, a concentrated region of yolk that, when viewed from above, both in unincubated eggs and in early stages of development, looks like a small white disc. Numerous observers, Malpighi and Harvey among them, frequently mistook this "white point" for either the embryo or the developing heart.

11 For a more detailed explanation of Wolff's theory, see Adelmann 1966, 4:1653–57.

12 Regarding Haller's authorship of these reviews, see Chapter 2, note 4.

13 Wolff's work on the intestines was cited by von Tredern, Tiedemann, Burdach, Oken, Döllinger, Pander, and von Baer, among others. See Adelmann 1966, 4:1679–1739, for the impact of Wolff's work on German embryology. On the history of the germ layer concept, see Oppenheimer 1967:256–94.

14 See Needham 1934, 1959 ed.:158, 185, 233; Adelmann 1966, 5:2236–42. On the microscopes and techniques that Haller used, see Mazzolini 1977: 220–24.

15 Wolff's "globules" were most likely optical artifacts. As Hall (1969, 2:186) has remarked, the history of the globular theory is largely the history of an illusion, one due to diffraction haloes caused by improper lighting techniques and to compound microscopes uncorrected for spherical and chromatic aberration.

Chapter 4. The philosophical debate: Newtonianism versus rationalism

1 See d'Irsay 1930; Guthke 1962, 1970, 1973, 1975; Sonntag 1971, 1974a, 1974b, 1975, 1977; and Toellner 1971, 1976.
2 See Lundsgaard-Hansen-von Fischer's (1959) catalog of Haller's published works, which contains over 300 main entries, a majority of which are books, and an additional 400 editions and translations. This list does not include Haller's ten- to twelve-thousand book reviews (see Chapter 2, note 4).
3 See Hochdoerfer 1932; Guthke 1962:17–19, 1975:174–92; and Toellner 1971:1–27, who provides a review of the literature on this subject.
4 Original text: "Genug, es ist ein Gott; es ruft es die Natur,/Der ganze Bau der Welt zeigt seiner Hände Spur."
5 Original text: "Der Mensch, vor dessen Wort sich soll die Erde bücken,/Ist ein Zusammenhang von eitel Meister-Stücken;/In ihm vereinigt sich der Körper Kunst und Pracht,/Kein Glied ist, das ihn nicht zum Herrn der Schöpfung macht."
6 On Haller's debate with Whytt, see Haller 1756–60; Whytt 1751, 1755; and French 1969. For Haller and La Mettrie, see Haller 1751d; Bergmann 1913; Vartanian 1949; Saussure 1949; Guthke 1962a; and Hintzsche 1968. See also Bastholm 1950.
7 See Richter 1972:60–68; Roe 1981. In his publications and diaries, Haller refers to Newton's *Opticks*, the *Principia*, the Leibniz–Clarke correspondence (Clarke 1715), 'sGravesande's *Physices elementa mathematica* (1720–21), and Henry Pemberton's *View of Sir Isaac Newton's Philosophy* (1728). See Haller 1746c and 1774–77, 1:621–22; Sigerist 1923:22–23; Guthke 1967:150; and Hintzsche and Balmer 1971:80, 91, 93.
8 See Jones 1925; Price 1926; Teeter 1928; and Richter 1972:57–111.
9 Original text: "Ein Newton übersteigt das Ziel erschaffner Geister,/Findt die Natur im Werk und scheint des Weltbaus Meister;/Er wiegt die innre Kraft, die sich im Körper regt,/Den einen sinken macht und den im Kreis bewegt,/Und schlägt die Tafeln auf der ewigen Gesetze,/Die Gott einmal gemacht, dass er sie nie verletze."
10 Original text:" . . . Er füllt die Welt mit Klärheit,/Er ist ein stäter Quell von unerkannter Wahrheit."
11 See, for example, Haller 1750a:xiii–xiv, xvii, xviii–xix; 1757–66, 4:533, 557; and 1774–77, 1:621–22.
12 I have altered this translation from "frame" to "feign" following Cohen 1971:241, n. 9.
13 I am reading "partibus" for "particulis" in the phrase "Theoriam autem, cur utravis proprietas aut in his partibus nulla sit, aut in aliis corporis humani particulis [*sic*] aliqua. . . ."
14 See also Haller's adoption of Newton's ideas on motion in his *Elementa physiologiae*, 1757–66, 4:557–58.
15 See also Oakley 1961; Kubrin 1967; McGuire 1968; Cohen 1969; Heimann and McGuire 1971; and Tamney 1979.
16 On Christian Wolff's philosophy, see Gurr 1959; Blackwell 1961; Burns 1966; Beck 1969; Corr 1972; and Frängsmyr 1975.
17 On Hoffmann and Stahl, see Lemoine 1864; Rather 1961; King 1964, 1969; and Duchesneau 1976.
18 Wolff's use of the verb *paterer* (imperfect subjunctive of *patior*) has engendered considerable controversy among Wolff scholars, who have frequently translated *paterer* in the sense of "allow" rather than "suffer from"

(see, for example, the German translation, Wolff 1896, no. 85:81; and the Russian translation, Wolff 1950:203). In my opinion, the "suffer from" translation is the correct one for two reasons. First, it is consistent with the other discussions of the soul in Wolff's dissertation (cited above). Second, in a later work of Wolff's titled *Von der eigenthümlichen und wesentlichen Kraft* (1789), Wolff, in complaining of Haller's earlier critique of his dissertation, remarked that "Haller has thus not been entirely correct . . . when in the judgment of my work he put forward the essential force as the main point, remembering all that properly belongs to this theory almost with no words and then deliberately mentioning that this thing is called by me the essential force, but which *I completely separated from the soul of the Stahlians*; about which indeed exactly so much as nothing was said" (p. 50 n.; italics added). Wolff's perception thirty years later was that he had clearly distinguished his essential force from the soul and from Stahl's ideas. That he was referring to the passage I am now discussing is borne out by the fact that nowhere else in the dissertation are Stahl or the "Stahlians" mentioned. (On the translation of this passage see also Mario Stenta's comments in Driesch 1911:379–83, n. 78; and Gaissinovitch 1961:253–55.)

19 On Haller's critique of Needham's views see Duchesneau 1979:80–85.

Chapter 5. Wolff's later work on variation and heredity

1 For a list of these materials, see Rajkov 1964:625–26 and Wolff 1973:304–5. See also von Baer 1847, for the earliest report of Wolff's manuscripts.

2 All citations will be to the Latin text in Wolff 1973.

3 In all, Wolff published thirty-six papers in the Academy of Sciences' journal during his years in St. Petersburg. Aside from the papers on monsters and on the formation of the intestines, these are mostly anatomical investigations, including several studies of the muscular fibers of the heart. For a complete list of Wolff's papers, see Rajkov 1964:623–25 and Wolff 1973:302–4.

4 See Gaissinovitch 1961:444–50; Rajkov 1964:605–18; and Lukina 1975:420–21.

5 Wolff refers to Ethiopians several times in his treatise in discussing variations in human traits, and at one point he cites a tract called "Ethiopian Anatomy," apparently written by himself (1973:227, § 75). Lukina, who has not been able to locate this paper among Wolff's manuscripts, suggests that it was written after 1772, when Wolff was sent the body of an African man for dissection (see Wolff 1973: 289, n. 21).

6 See Stearn 1957:156–61; Glass 1959c: and Larson 1971:94–121.

7 See Roger 1963:567–82; Farber 1972; and Bowler 1973.

8 See Gaissinovitch 1961:444–50 and Rajkov 1964:600–605. For a Russian translation of the "Distributio operis," see Gaissinovitch 1961:526–33. All references will be to Gaissinovitch's translation.

9 See Gaissinovitch 1961:426, n. 682; 1978:525; Rajkov 1964:619–20; and Lukina 1975:417–18. The Latin text of this section is included in Wolff 1973:288–89.

10 I am not arguing that a belief in rationalism necessarily entails support for epigenesis. Both Leibniz and Christian Wolff, for example, supported theories of spermaticist preformation (see Christian Wolff 1723, § 444; Leibniz 1951:109, 195–97, 276, 526–27, 548–49, 550; and Roger 1968). What I do wish to claim is that, in Caspar Friedrich Wolff's case, his belief

in rationalism, together with his philosophy of biology, provided the basis on which an epigenetic theory of development could be built.

Chapter 6. Epilogue: the old and the new

1 On the Harvey–Descartes controversy, see Passmore 1958; Mendelsohn 1964; and Toellner 1972. Regarding spontaneous generation, see Farley 1977 and, on the Pasteur–Pouchet debate, Farley and Geison 1974. Concerning Stahl and Hoffmann, see Chapter 4, note 17. For the Lawrence–Abernethy controversy, see Temkin 1963 and Goodfield-Toulmin 1969. On Roux and Driesch, see Churchill 1969, 1973; and Allen 1975.

2 On incommensurability and related philosophical and social aspects of scientific controversy, see, for example, Hanson 1965; Scheffler 1967; Kuhn 1970; Lakatos and Musgrave 1970; Suppe 1974; Barnes 1974; and Mulkay 1979.

3 See Temkin 1950; Larson 1979; and Lenoir 1980.

4 Not all of the German embryologists I am referring to accepted the recapitulation theory. Von Baer, for one, was critical of the views of Meckel and other proponents of recapitulation, substituting his own theory of divergent development – that among the four basic types of organisms embryonic development proceeds from the general to the specific, that is, for example, from a generalized vertebrate form to a special kind of vertebrate (see Oppenheimer 1967:221–55; Ospovat 1976; and Gould 1977). Meckel and the other recapitulationists believed that the embryo proceeds through the adult forms of lower organisms (in an ideal sense) during its developmental sequence. It should also be noted that these early proponents of the biogenetic law believed that ontogeny recapitulates phylogeny in an ideal, or transcendental, sense rather than in an evolutionary one. The biogenetic law received its first evolutionary formulation later in the nineteenth century in the work of Fritz Müller and Ernst Haeckel (see Gould 1977).

5 See, for example, Oscar Hertwig's tract, *The Biological Problem of To-day: Preformation or Epigenesis?* (1896), which opposed August Weismann's germ plasm theory of inheritance. Preformation–epigenesis controversies also arose in the early twentieth century in association with debates over Mendelian genetics and over the chromosome theory. See Churchill 1968, 1970b; Allen 1975; and Baxter 1976.

6 See Kohlbrugge 1913:218, n. 14; Temkin 1950; and Herrlinger 1966:22–28. Herrlinger notes that Wolff's writings were eclipsed by the popularity of Blumenbach's *Über den Bildungstrieb und das Zeugungsgeschäfte* (1781 and later editions).

7 See Temkin 1950; Oppenheimer 1967:136–47; Gasking 1967; Gould 1977; and Lenoir 1978.

Appendix B. Wolff's letters to Haller

1 Wolff had probably received from Haller some of the publications of Johann Adrian Theodor Sproegel, one of Haller's former students, who completed his medical degree at Göttingen in 1753.

2 By the end of 1761, three volumes of Haller's *Elementa physiologiae* had been published.

3 Bonnet had defended preformation and attacked epigenesis in his *Considérations sur les corps organisés* (1762) without mentioning Wolff or his theory of epigenesis. Wolff interpreted this as an intentional snub and

responded by attacking and ridiculing Bonnet's views in his *Theorie von der Generation* (1764). Although Haller apparently maintained, in his reply to this letter, that Bonnet had indeed not read Wolff's dissertation and had only heard of it through Haller's letters, Wolff was not satisfied, as can be seen in his subsequent letter (Letter IV).

4 See Wolff's paper, "De foramine ovali, ejusque usu, in dirigendo motu sanguinis. Observationes novae," *Novi Commentarii Academiae Scientiarum Imperialis Petropolitanae*, 20(1775):357–430. After this volume, the Academy's journal was titled *Acta Academiae Scientiarum Imperialis Petropolitanae*.

5 The foramen ovale is an opening in the septum between the two auricles of the mammalian fetal heart. Before birth, the fetal lungs do not operate, and the fetus receives oxygenated blood through the umbilical vein. This blood enters the right auricle of the heart through the inferior vena cava. The foramen ovale allows the blood to bypass the fetal lungs by sending it directly from the right auricle into the left auricle, rather than into the right ventricle and thence to the pulmonary circuit. The foramen ovale closes shortly after birth when the lungs begin to function and normal circulation is established. Analogous openings between the two auricles exist in the fetal chick and in the embryos of other animals with four-chambered hearts.

6 Wolff uses the word "sinus" to refer to the entire region of the right or left auricle. He reserves the term "auricle" for an ear-shaped portion of the auricle that projects out over the surface of the heart. Modern anatomists often distinguish between "atrium" and "auricle" in the same manner.

7 See Winslow 1717, 1725a, 1725b; and Rouhault 1728. An earlier controversy over the function of the foramen ovale also took place at the turn of the century between Jean Méry and Joseph-Guichard Duverney. See the volumes of the *Histoire de l'Académie Royal des Sciences* for the years 1699, 1701, and 1703.

8 Both Bonnet and Haller, along with several other notable figures, were made foreign members of the St. Petersburg Academy of Sciences in honor of its fiftieth anniversary. See *Procès-verbaux des séances de l'académie impériale des sciences* (1897-1911), 3:272–75, 298.

9 Wolff's account of fetal circulation and of the role played by the foramen ovale was erroneous in several respects. Although his illustrations of the embryonic heart are quite accurate (see Figure 21), Wolff was incorrect in claiming that the inferior vena cava sends blood separately to the right and left auricles. The blood that enters the right auricle from the inferior vena cava passes (for the most part) directly through the foramen ovale into the left auricle (and then to the left ventricle). Although some blood does indeed pass from the right auricle to the right ventricle, this is blood that has entered the right auricle from the superior vena cava. The two streams of blood (from the inferior vena cava and the superior vena cava) remain quite separate in the right auricle (see Patten 1968). What may have confused Wolff is the fact that, in the fetal heart, the valve of the inferior vena cava (the Eustachian valve) is continuous with the border of the foramen ovale. Thus, it would be possible to view the inferior vena cava as continuing directly into the left auricle.

Haller was apparently unimpressed with Wolff's anatomical investigation on the foramen ovale. As he wrote Johannes Gessner after receiving Wolff's paper: "He says nothing in all these pages except that the foramen ovale is a canal that has two openings, but this had been known already for a long time" (Sigerist 1923:535; letter of 19 April 1777).

Bibliography

Unpublished Materials

In addition to the manuscript sources listed below, see Wolff 1973 for a recent publication of Wolff's manuscript treatise, "Objecta meditationum pro theoria monstrorum." For Wolff's letters to Haller, see also Haller 1773–75; Schuster 1941; Gaissinovitch 1961: 510–25; and Appendix B of this book.

"Briefe an Albrecht v. Haller." MSS Hist. Helv. XVIII. Burgerbibliothek, Bern. Contains originals of letters to Haller, including those from Bonnet and Wolff (cited in text as "Haller MSS").

"Lettres en original de M. le Baron de Haller à Charles Bonnet." 9 vols. MS Bonnet 44–52 (1754–77). Bibliothèque Publique et Universitaire, Geneva. Contains Haller's letters to Bonnet (cited in text as "Bonnet MSS").

"Kopie der Correspondenz Charles Bonnet–Albrecht von Haller." Haller 91.1 (1754–63), 91.2 (1763–66). Burgerbibliothek, Bern.

"Observationes anatomicae Bernenses." 3 vols. MSS Haller 16 (1753–57), 17 (1757–62), 18 (1762–[65]). Burgerbibliothek, Bern.
Contains Haller's observations on incubated chicken eggs.

Published Sources

Adelmann, Howard B.
1966 *Marcello Malpighi and the Evolution of Embryology*. 5 vols. Ithaca, New York: Cornell University Press.

Allen, Garland
1975 *Life Science in the Twentieth Century*. New York & London: John Wiley & Sons. Cambridge: Cambridge University Press, 1978.

Anderson, Lorin
1976 "Charles Bonnet's Taxonomy and Chain of Being." *Journal of the History of Ideas*, 37:45–58.

Arber, Agnes
1946 "Goethe's Botany. *The Metamorphosis of Plants* (1790) and Tobler's *Ode to Nature* (1782)." *Chronica Botanica*, 10:63–126.

Aulie, Richard P.
1961 "Caspar Friedrich Wolff and his 'Theoria Generationis,' 1759." *Journal of the History of Medicine*, 16:124–44.

Baer, Karl Ernst von
1827 *De ovi mammalium et hominis genesi epistola*. Leipzig: L. Voss.
1828–37 *Über Entwickelungsgeschichte der Thiere*. 2 vols. Part 1, 1828; part 2, 1837. Königsberg: Bornträger.
1847 "Über den litterärischen Nachlass von Caspar Friedrich

Wolff, ehemaligen Mitgliede der Akademie der Wissenschaften zu St. Petersburg." *Bulletin de la classe physico-mathématique de l'Académie Impériale des Sciences de Saint-Pétersbourg*, 5:129–60.

1866 *Nachrichten über Leben und Schriften*. St. Petersburg: H. Schmitzdorff.

1956 "On the Genesis of the Ovum of Mammals and of Man." Translated by Charles Donald O'Malley. Introduction by I. Bernard Cohen. *Isis*, 47:117–53. Translation of von Baer 1827.

Baker, John R.
1952 *Abraham Trembley of Geneva: Scientist and Philosopher, 1710–1784*. London: Edward Arnold.

Barnes, Barry
1974 *Scientific Knowledge and Sociological Theory*. London: Routledge & Kegan Paul.

Bastholm, Eyvind
1950 *The History of Muscle Physiology from the Natural Philosophers to Albrecht von Haller*. Acta Historica Scientiarum Naturalium et Medicinalium, no. 7. Copenhagen: E. Munksgaard.

Baxter, Alice Levine
1976 "Edmund B. Wilson as a Preformationist: Some Reasons for His Acceptance of the Chromosome Theory." *Journal of the History of Biology*, 9:29–57.

Beck, Lewis White
1969 *Early German Philosophy: Kant and his Predecessors*. Cambridge: Harvard University Press.

Beer, Rüdiger Robert
1947 *Der grosse Haller*. Säckingen: Hermann Stratz.

Belloni, Luigi
1971 "Embryological Drawings concerning his *Theorie von der Generation* sent by Caspar Friedrich Wolff to Albrecht von Haller in 1764." *Journal of the History of Medicine*, 26:205–8.

Bentley, Richard
1693 *The Folly and Unreasonableness of Atheism, Demonstrated from the Advantage and Pleasure of a Religious Life, the Faculties of Human Souls, the Structure of Animate Bodies, & the Origin and Frame of the World: In Eight Sermons*. London: H. Mortlock.

Bergmann, Ernst
1913 *Die Satiren des Herrn Maschine: Ein Beitrag zur Philosophie- und Kulturgeschichte des 18. Jahrhunderts*. Leipzig: E. Wiegandt.

Bilikiewicz, Tadeusz
1932 *Die Embryologie im Zeitalter des Barock und des Rokoko*. Arbeiten des Instituts für Geschichte der Medizin an der Universität Leipzig, no. 2. Leipzig: Georg Thieme.

Blackwell, Richard J.
1961 "The Structure of Wolffian Philosophy." *The Modern Schoolman*, 38:203–18.

Blumenbach, Johann Friedrich
1781 *Über den Bildungstrieb und das Zeugungsgeschäfte*. Göttingen: Johann Christian Dieterich. 2nd ed., 1789.

Blumenbach, Johann Friedrich, and Born, Carl Friedrich
1789 *Zwo Abhandlungen über die Nutritionskraft welche von der Kayserlichen Academie der Wissenschaften in St. Petersburg den Preis getheilt erhalten haben. Nebst einer fernern Erläuterung eben derselben Materie von C. F. Wolff.* St. Petersburg: Kayserlichen Academie der Wissenschaften.
Bodemann, Eduard, ed.
1885 *Von und über Albrecht von Haller: Ungedruckte Briefe und Gedichte Hallers sowie ungedruckte Briefe und Notizen über denselben.* Hanover: Carl Meyer.
Bodemer, Charles W.
1964 "Regeneration and the Decline of Preformationism in Eighteenth Century Embryology." *Bulletin of the History of Medicine*, 38:20–31.
1968 "Embryological Thought in Seventeenth Century England." In *Medical Investigations in Seventeenth Century England*, pp. 3–25. University of California, Los Angeles: William Andrews Clark Memorial Library.
Boerhaave, Herman
1708 *Institutiones medicae, in usus annuae exercitationis domesticos.* Leiden: J. van der Linden. Many later editions. See also Haller 1739–44.
Bonnet, Charles
1745 *Traité d'insectologie, ou observations sur les pucerons et sur quelques espèces de vers d'eau douce, qui coupé par morceaux, deviennent autant d'animaux complets.* Paris: Durand.
1762 *Considérations sur les corps organisés, où l'on traite de leur origine, de leur développement, de leur réproduction, etc.* 2 vols. Amsterdam: M. M. Rey.
1779–83 *Oeuvres d'histoire naturelle et de philosophie.* 18 vols. Neuchatel: S. Fauche.
Bonnet, Georges
1929 *Charles Bonnet (1720–1793).* Paris: M. Lac.
Bowler, Peter J.
1971 "Preformation and Pre-existence in the Seventeenth Century: A Brief Analysis." *Journal of the History of Biology*, 4:221–44.
1973 "Bonnet and Buffon: Theories of Generation and the Problem of Species." *Journal of the History of Biology*, 6:259–81.
1975 "The Changing Meaning of 'Evolution.'" *Journal of the History of Ideas*, 36:95–114.
Bradbury, Saville
1967 *The Evolution of the Microscope.* Oxford: Pergamon Press.
Buffon, George Louis Leclerc, Comte de
1749–89 *Histoire naturelle, générale et particulière, avec la description du cabinet du roy.* 31 vols. Paris: L'Imprimerie Royale.
1750–72 *Allgemeine Historie der Natur nach allen ihren besondern Theilen abgehandelt.* 8 vols. Hamburg & Leipzig: G. C. Grund and A. H. Holle. Translation of the first 8 volumes of Buffon 1749–89.
Burns, John V.
1966 *Dynamism in the Cosmology of Christian Wolff: A Study in Precritical Rationalism.* New York: Exposition Press.

Castellani, Carlo
1965 *La storia della generazione, idee e teorie dal diciassettesimo al diciottesimo secolo*. Milan: Longanesi.
1969–70 "L'origine degli infusori nella polemica Needham, Spallanzani, Bonnet." *Episteme*, 3:214–41, 4:19–36.
1971 Ed. *Lettres à M. L'abbé Spallanzani de Charles Bonnet*. Milan: Episteme Editrice.
1972 "The Problem of Generation in Bonnet and Buffon: A Critical Comparison." In *Science, Medicine, and Society in the Renaissance: Essays to Honor Walter Pagel*, 2:265–88. Edited by Allen G. Debus. 2 vols. New York: Science History Publications.

Churchill, Frederick B.
1968 "August Weismann and a Break from Tradition." *Journal of the History of Biology*, 1:91–112.
1969 "From Machine-Theory to Entelechy: Two Studies in Developmental Teleology." *Journal of the History of Biology*, 2:165–85.
1970a "The History of Embryology as Intellectual History." *Journal of the History of Biology*, 3:155–81.
1970b "Hertwig, Weismann, and the Meaning of Reduction Division circa 1890." *Isis*, 61:429–57.
1973 "Chabry, Roux, and the Experimental Method in Nineteenth-Century Embryology." In *Foundations of Scientific Method: The Nineteenth Century*, pp. 161–205. Edited by Ronald N. Giere and Richard S. Westfall. Bloomington, Ind.: Indiana University Press.
1976 "Rudolf Virchow and the Pathologist's Criteria for the Inheritance of Acquired Characteristics." *Journal of the History of Medicine*, 31:117–48.

Clarke, Samuel
1715 *A Collection of Papers, which Passed between the Late Learned Mr. Leibnitz, and Dr. Clarke, in the Years 1715 and 1716*. London: J. Knapton.

Cohen, I. Bernard
1969 "Isaac Newton's *Principia*, the Scriptures, and the Divine Providence." In *Philosophy, Science, and Method: Essays in Honor of Ernest Nagel*, pp. 523–48. Edited by Sidney Morgenbesser, Patrick Suppes, and Morton White. New York: St. Martin's Press.
1971 *Introduction to Newton's "Principia."* Cambridge, Mass.: Harvard University Press. Cambridge, Eng.: Cambridge University Press.

Cole, F. J.
1930 *Early Theories of Sexual Generation*. Oxford: Clarendon Press.

Condorcet, Marie Jean Antoine Nicolas Caritat, Marquis de
1780 "Éloge de M. de Haller." *Histoire de l'Académie Royale des Sciences, année 1777*, pp. 210–59. Paris: L'Imprimerie Royale.

Corcos, Alain F.
1971 "Fontenelle and the Problem of Generation in the Eighteenth Century." *Journal of the History of Biology*, 4:363–72.

Corr, Charles A.
1972 "Christian Wolff's Treatment of Scientific Discovery." *Journal of the History of Philosophy*, 10:323–34.

Descartes, René
1644 *Principia philosophiae*. Reprinted in *Oeuvres* (1964–74), vol. 8, pt. 1.
1664 *L'Homme et un traitté de la formation du foetus*. With remarks by Louis de la Forge. Paris: Charles Angot.
1964–74 *Oeuvres*. Edited by Charles Adam and Paul Tannery. 11 vols. Paris: Librairie Philosophique J. Vrin.
1972 *Treatise of Man*. French–English edition, with commentary by Thomas Steele Hall. Cambridge: Harvard University Press.

Diderot, Denis
1754 *Pensées sur l'interpretation de la nature*. Paris: n.p.
1769 *Le Rêve de d'Alembert*. Ed. Paul Vernière. Paris: Didier, 1951.

d'Irsay, Stephen
1930 *Albrecht von Haller: Eine Studie zur Geistesgeschichte der Aufklärung*. Arbeiten des Instituts für Geschichte der Medizin an der Universität Leipzig, no. 1. Leipzig: Georg Thieme.

Dolman, Claude E.
1975 "Spallanzani, Lazzaro." *Dictionary of Scientific Biography*, 12: 553–67. New York: Charles Scribner's Sons.

Driesch, Hans
1911 *Il vitalismo, storia e dottrina*. Translated and with notes added by Mario Stenta. Milan: Remo Sandron.

Duchesneau, François
1976 "G. E. Stahl: Antimécanisme et physiologie." *Archives internationales d'histoire des sciences*, 26:3–26.
1979 "Haller et les théories de Buffon et C. F. Wolff sur l'épigenèse." *History and Philosophy of the Life Sciences*, 1:65–100. Pubblicazioni della stazione zoologica di Napoli, Section II.

Fambach, Oscar
1976 *Die Mitarbeiter der Göttingischen Gelehrten Anzeigen 1769–1836*. Tübingen: Universitätsbibliothek.

Farber, Paul L.
1972 "Buffon and the Concept of Species." *Journal of the History of Biology*, 5:259–84.

Farley, John
1977 *The Spontaneous Generation Controversy from Descartes to Oparin*. Baltimore & London: Johns Hopkins University Press.

Farley, John, and Geison, Gerald L.
1974 "Science, Politics and Spontaneous Generation in Nineteenth-Century France: The Pasteur–Pouchet Debate." *Bulletin of the History of Medicine*, 48:161–98.

Fontenelle, Bernard Le Bovier de
1683 *Lettres diverses de M. le Chevalier d'Her****. Reprinted as *Lettres galantes* in *Oeuvres*, vol. 1. 11 vols. Paris: Libraires Associés, 1766.

Foote, Edward T.
1969 "Harvey: Spontaneous Generation and the Egg." *Annals of Science*, 25:139–63.

Frängsmyr, Tore
1975 "Christian Wolff's Mathematical Method and Its Impact on the Eighteenth Century." *Journal of the History of Ideas*, 36: 653–68.

French, R. K.
1969 *Robert Whytt, the Soul, and Medicine*. London: Wellcome Institute of the History of Medicine.

Gaissinovitch, A. E.
1956–57 "Notizen von C. F. Wolff über die Bermerkungen der Opponenten zu seiner Dissertation." *Wissenschaftliche Zeitschrift der Friedrich-Schiller-Universität Jena*, 6:121–24.
1961 *K. F. Vol'f i uchenie o razvitii organizmov (v svjazi s obshchej èvoljuciej nauchnogo mirovozzrenija)* [K. F. Wolff and his doctrine of the development of organisms (in connection with the general evolution of his scientific world view)]. Moscow: Izdatel'stvo akademii nauk SSSR.
1968 "Le Rôle du Newtonianisme dans la renaissance des idées épigénétiques en embryologie du XVIIIe siècle." *Actes du XIe Congrès International d'Histoire des Sciences 1965*, 5:105–10. Varsovie.
1978 "Wolff, Caspar Friedrich." *Dictionary of Scientific Biography*. 15:524–26. New York: Charles Scribner's Sons.

Garden, George
1691 "A Discourse concerning the Modern Theory of Generation." *Philosophical Transactions of the Royal Society of London*, 17:474–83.

Gasking, Elizabeth
1967 *Investigations into Generation, 1651–1828*. Baltimore: Johns Hopkins Press. London: Hutchinson Publishing Group.

Glass, Bentley
1959a "The Germination of the Idea of Biological Species." In *Forerunners of Darwin: 1745–1859*, pp. 30–48. Edited by Bentley Glass, Owsei Temkin, and William L. Straus, Jr. Baltimore: Johns Hopkins Press.
1959b "Maupertuis, Pioneer of Genetics and Evolution." In *Forerunners of Darwin: 1745–1859*, pp. 51–83. Edited by Bentley Glass, Owsei Temkin, and William L. Straus, Jr. Baltimore: Johns Hopkins Press.
1959c "Heredity and Variation in the Eighteenth Century Concept of the Species." In *Forerunners of Darwin: 1745–1859*, pp. 144–72. Edited by Bentley Glass, Owsei Temkin, and William L. Straus, Jr. Baltimore: Johns Hopkins Press.
1974 "Maupertuis, Pierre Louis Moreau de." *Dictionary of Scientific Biography*, 9:186–89. New York: Charles Scribner's Sons.

Goethe, Johann Wolfgang von
1817–24 *Zur Morphologie*. 2 vols. Stuttgart & Tübingen: J. C. Cotta'schen.

Goodfield-Toulmin, June
1969 "Some Aspects of English Physiology: 1780–1840." *Journal of the History of Biology*, 2:283–320.

Gould, Stephen Jay
1977 *Ontogeny and Phylogeny*. Cambridge, Mass. & London: Harvard University Press.

Graaf, Regnier de
1672 *De mulierum organis generationi inservientibus tractatus novus*. Leiden: Hack.

'sGravesande, Willem Jacob van
1720–21 *Physices elementa mathematica, experimentis confirmata. Sive introductio ad philosophiam Newtonianam.* Leiden: Petrum van der Aa.

Gurr, John Edwin
1959 *The Principle of Sufficient Reason in Some Scholastic Systems, 1750–1900.* Milwaukee: Marquette University Press.

Guthke, Karl S.
1962a "Haller, La Mettrie und die anonyme Schrift *L'Homme plus que machine.*" *Études Germaniques,* 17:137–43.
1962b *Haller und die Literatur.* Arbeiten aus der Niedersächsischen Staats- und Universitätsbibliothek Göttingen, vol. 4. Göttingen: Vandenhoeck & Ruprecht.
1967 "Zur Religionsphilosophie des jungen Albrecht von Haller." *Colloquia Germanica,* 1:142–55.
1970 Ed. *Hallers Literaturkritik.* Freies Deutsches Hochstift Reihe der Schriften, no. 21. Tübingen: Max Niemeyer.
1973 "Haller und die 'Bibliothèque raisonnée.'" *Jahrbuch des Freien Deutschen Hochstifts,* pp. 1–13.
1975 *Literarisches Leben im achtzehnten Jahrhundert in Deutschland und in der Schweiz.* Bern & Munich: Francke.

Guyénot, Émile
1957 *Les Sciences de la vie aux XVIIe et XVIIIe siècles. L'Idée d'évolution.* Paris: Éditions Albin Michel.

Hall, Thomas S.
1968 "On Biological Analogs of Newtonian Paradigms." *Philosophy of Science,* 35:6–27.
1969 *History of General Physiology, 600 B.C. to A.D. 1900.* 2 vols. Chicago & London: University of Chicago Press.
1970 "Descartes' Physiological Method: Position, Principles, Examples." *Journal of the History of Biology,* 3:53–79.

Haller, Albrecht von
1739 *Descriptio foetus bicipitis ad pectora connati ubi in causas monstrorum ex principiis anatomicis inquiritur.* Hannover: B. Nic. Foerster.
1739–44 Ed. *Praelectiones academicae in proprias institutiones rei medicae,* by Herman Boerhaave. Notes added by Albrecht von Haller. 6 vols. in 7. Göttingen: A. Vandenhoeck.
1743–54 *Icones anatomicae quibus praecipuae aliquae partes corporis humani delineatae proponuntur et arteriarum potissium historia contineatur.* Fasc. I-VIII. Göttingen: A. Vandenhoeck.
1746a Review of *Traité d'insectologie,* by Charles Bonnet. *Bibliothèque raisonnée,* 36:179–92. Haller's authorship: listed in Guthke 1973 and 1975:348–53; reprinted in translation in Haller 1772b.
1746b Review of *Philosophia rationalis, quae logico vulgo dicitur, multum aucta et emendata. Paullo uberioris in universam philosophiam introductionis Pars I,* by Samuel Christian Hollmann. *Bibliothèque raisonnée,* 37:355–65. Haller's authorship: listed in Guthke 1973 and 1975:348–53; reprinted in translation in Haller 1772b.
1746c Review of *Philosophiae naturalis principia mathematica,* by Isaac

Newton, commentary by P. P. Thomas Le Seur and François Jaquier, 4 vols. (1739–42). *Bibliothèque raisonnée*, 37:54–61. Haller's authorship: listed in Guthke 1973 and 1975:348–53.

1747 *Primae lineae physiologiae in usum praelectionum academicarum*. Göttingen: A. Vandenhoeck.

1750a Preface to *Allgemeine Historie der Natur*, vol. 1, by Buffon. Hamburg & Leipzig: G. C. Grund and A. H. Holle. Reprinted in Haller 1756 and 1772b.

1750b Review of *Histoire naturelle*, vol. 1, by Buffon. *Bibliothèque raisonnée*, 45:243–63. Haller's authorship: listed in Guthke 1973 and 1975:348–53.

1751a *Réflexions sur le système de la génération de M. de Buffon*. Geneva: Barrillot.

1751b Review of *Histoire naturelle*, vol. 2, by Buffon. *Bibliothèque raisonnée*, 46:68–88. Haller's authorship: listed in Guthke 1973 and 1975:348–53.

1751c *Primae lineae physiologiae in usum praelectionum academicarum*. 2nd ed. Göttingen: A. Vandenhoeck.

1751d *Lettre de M. de Haller à M. de Maupertuis, sur une brochure de M. de la Mettrie avec la réponse de M. de Maupertuis*. Göttingen: n.p.

1752a Preface to *Allgemeine Historie der Natur*, vol. 2, by Buffon. Hamburg & Leipzig: G. C. Grund and A. H. Holle. Reprinted in Haller 1756 and 1772b.

1752b "De partibus corporis humani sensilibus et irritabilibus." *Commentarii Societatis Regiae Scientiarum Gottingensis*, 2:114–58. Published 1753.

1754 *Dr. Albert Haller's Physiology, being a Course of Lectures upon the Visceral Anatomy and Vital Oeconomy of Human Bodies*. 2 vols. London: W. Innys & J. Richardson. Translation of Haller 1751c.

1756 *Sammlung kleiner Hallerischer Schriften*. Bern: E. Haller.

1756–60 *Mémoires sur la nature sensible et irritable des parties du corps animal*. 4 vols. Vol 1, Lausanne: M. M. Bousquet; vols. 2, 3, 4, Lausanne: S. d'Arnay.

1757–66 *Elementa physiologiae corporis humani*. 8 vols. Lausanne: M. M. Bousquet, S. d'Arnay, F. Grasset. Bern: Societas Typographica.

1758a *Sur la formation du coeur dans le poulet; sur l'oeil, sur la structure du jaune, etc.* 2 vols. Lausanne: M. M. Bousquet.

1758b *Deux Mémoires sur la formation des os, fondés sur des expériences*. Lausanne: M. M. Bousquet.

1760 Review of *Theoria generationis*, by Caspar Friedrich Wolff. *Göttingische Anzeigen von gelehrten Sachen*, pp. 1226–31. Haller's authorship: "H" in Bern copy.

1762–68 *Opera minora, emendata, aucta, et renovata*. 3 vols. Lausanne: F. Grasset. Volumes 2 and 3 titled *Operum anatomici argumenti minorum*.

1765a Review of *Theorie von der Generation*, by Caspar Friedrich Wolff. *Göttingische Anzeigen von gelehrten Sachen*, pp. 549–52. Haller's authorship: "H" in Bern copy.

1765b *Primae lineae physiologiae in usum praelectionum academicarum*. 3rd ed. Göttingen: A. Vandenhoeck.

1770 Review of *Novi Commentarii Academiae Scientiarum Imperialis Petropolitanae*, vol. 12 (containing Parts 1 and 2 of Wolff's "De formatione intestinorum," 1766–67). *Zugabe zu den Göttingischen Anzeigen von gelehrten Sachen*, pp. 377–81. Haller's authorship: "H" in Bern copy.
1771 Review of *Novi Commentarii Academiae Scientiarum Imperialis Petropolitanae*, vol. 13 (containing Part 3 of Wolff's "De formatione intestinorum," 1768). *Zugabe zu den Göttingischen Anzeigen von gelehrten Sachen*, pp. 414–16. Haller's authorship: "H" in Bern copy.
1772a *Briefe über die wichtigsten Wahrheiten der Offenbarung*. Bern: Neue Buchhandlung.
1772b *Sammlung kleiner Hallerischer Schriften*. 2nd ed. 3 vols. Bern: E. Haller.
1773 Review of *Oeuvres de M. Thomas*, vol. 4 (containing Thomas's "Éloge de Descartes"). *Zugabe zu den Göttingischen Anzeigen von gelehrten Sachen*, pp. 371–74. Haller's authorship: "H" in Bern copy.
1773–75 *Epistolarum ab eruditis viris ad Alb. Hallerum scriptarum. Pars I. Latinae*. 6 vols. Bern: Societas Typographica.
1774 *La Génération, ou exposition des phénomènes relatifs à cette fonction naturelle*. 2 vols. Paris: Des Ventes de la Doué. Translation of Haller 1757–66, vols. 7(part) and 8.
1774–77 *Bibliotheca anatomica qua scripta ad anatomen et physiologiam*. 2 vols. Zürich: Orell, Gessner, Fuessli & Socc.
1775–77 *Briefe über einige Einwürfe nochlebender Freygeister wieder die Offenbarung*. 3 vols. Bern: Typographische Gesellschaft.
1777a "Foetus." *Supplément à l'Encyclopédie*. 3:64–73. 4 vols. Amsterdam: M. M. Rey.
1777b "Génération." *Supplément à l'Encyclopédie*, 3:196–98. 4 vols. Amsterdam: M. M. Rey.
1777c "Oeconomie animale." *Supplément à l'Encyclopédie*. 4:104–5. 4 vols. Amsterdam: M. M. Rey.
1777d "Oeuf." *Supplément à l'Encyclopédie*, 4:120–23. 4 vols. Amsterdam: M. M. Rey.
1777e "Physiologie."*Supplément à l'Encyclopédie*, 4:344–65. 4 vols. Amsterdam: M. M. Rey.
1803 *First Lines of Physiology*. Troy, New York: Obadiah Penniman. Translation of Haller 1765b.
1936 "A Dissertation on the Sensible and Irritable Parts of Animals." Introduction by Owsei Temkin. *Bulletin of the History of Medicine*, 4:651–99. Based on a 1755 translation of Haller 1752.
1966 *First Lines of Physiology*. Facsimile reprint of the 1786 edition with an introduction by Lester S. King. The Sources of Science, no. 32. New York & London: Johnson Reprint Corporation. Translation of Haller 1765b.

Hamilton, Howard L., ed.
1952 *Lillie's Development of the Chick: An Introduction to Embryology*. New York: Holt, Rinehart & Winston.

Hanson, Norwood Russell
1965 *Patterns of Discovery*. Cambridge: Cambridge University Press.

Harvey, William
1651 *Exercitationes de generatione animalium*. London: O. Pulleyn.
1847 *The Works of William Harvey*. Translated by Robert Willis. London: Sydenham Society. Facsimile reprint ed., New York & London: Johnson Reprint Corporation, 1965.
Heimann, P. E.
1973 "Newtonian Natural Philosophy and the Scientific Revolution." *History of Science*, 11:1–7.
Heimann, P. E., and McGuire, J. E.
1971 "Newtonian Forces and Lockean Powers: Concepts of Matter in Eighteenth-Century Thought." *Historical Studies in the Physical Sciences*, 3:233–306.
Herrlinger, Robert
1959 "C. F. Wolffs 'Theoria generationis' (1759), Die Geschichte einer epochemachenden Dissertation." *Zeitscrift für Anatomie und Entwicklungsgeschichte*, 121:245–70.
1966 Introduction to *Theorie von der Generation in zwei Abhandlungen erklärt und bewiesen; Theoria generationis*, by Caspar Friedrich Wolff. Facsimile reprint ed. Hildesheim: Georg Olms.
Hertwig, Oscar
1896 *The Biological Problem of To-day: Preformation or Epigenesis? The Basis of a Theory of Organic Development*. Translated by P. Chalmers Mitchell. London: W. Heinemann.
Hertwig, Paula
1952 "Caspar Friedrich Wolff und Wilhelm Roux in ihrer Bedeutung für entwicklungsgeschichtliche Forschung (Halle, 1759 und 1895–1924)." *450 Jahre Martin-Luther-Universität Halle-Wittenberg*, 2:515–23. 3 vols. Halle-Wittenberg: n.p.
Hintzsche, Erich
1959 "Einige kritische Bemerkungen zur Bio- und Ergographie Albrecht von Hallers." *Gesnerus*, 16:1–15.
1964 Ed. *Albrecht von Haller, Giambattista Morgagni. Briefwechsel 1745–1768*. Bern: Hans Huber.
1965 Ed. *Albrecht von Haller, Ignazio Somis. Briefwechsel 1754–1777*. Bern: Hans Huber.
1966a Ed. *Albrecht von Haller, Marc Antonio Caldani. Briefwechsel 1756–1776*. Bern: Hans Huber.
1966b "Albrecht von Hallers Tätigkeit als Enzyklopädist." *Clio Medica*, 1:235–54.
1968 "Neue Funde zum Thema: L'homme machine und Albrecht Haller." *Gesnerus*, 25:135–66.
1972 "Haller, (Victor) Albrecht von." *Dictionary of Scientific Biography*, 6:61–67. New York: Charles Scribner's Sons.
1977 Ed. *Albrecht von Hallers Briefe an Auguste Tissot*. Bern: Hans Huber.
Hintzsche, Erich, and Balmer, Heinz, eds.
1971 *Albrecht Hallers Tagebücher seiner Reisen nach Deutschland, Holland und England 1723–1727*. Berner Beiträge zur Geschichte der Medizin und der Naturwissenschaften, n.s., vol. 4. Bern: Hans Huber.
Hirzel, Ludwig, ed.
1882 *Albrecht von Hallers Gedichte*. Frauenfeld: J. Huber.

Histoire de l'Académie Royale des Sciences, année 1741
1744 "Animaux coupés et partagés en plusieurs parties, et qui se reproduisent tout entiers dans chacune," pp. 33–35. Paris: L'Imprimerie Royale.

Hochdoerfer, Margarete
1932 *The Conflict between the Religious and the Scientific Views of Albrecht von Haller (1708–1777).* University of Nebraska Studies in Language, Literature, and Criticism, no. 12. Lincoln: n.p.

Hoffmann, Friedrich
1695 *Fundamenta medicinae ex principiis naturae mechanicis in usum philiatrorum succincte proposita.* Halle: S. J. Hübner.
1746 *Commentarius de differentia inter ejus doctrinam medico-mechanicam, et Georgii Ernesti Stahlii medico-organicam.* Frankfurt: F. Varrentrapp.
1971 *Fundamenta Medicinae.* Translated and with an introduction by Lester S. King. London: Macdonald. New York: Science History Publications. Translation of Hoffmann 1695.

Jacob, François
1973 *The Logic of Life: A History of Heredity.* Translated by Betty E. Spillmann. New York: Pantheon Books. London: Allen Lane.

Jenny, Heinrich Ernst
1902 *Haller als Philosoph, ein Versuch.* Basel: Basler Druck- und Verlags- Anstalt.

Jones, Howard Mumford
1925 "Albrecht von Haller and English Philosophy."*Publications of the Modern Language Association,* 40:103–27.

Kant, Immanuel
1790 *Kritik der Urtheilskraft.* Berlin: Lagarde & Friederich.
1966 *Critique of Judgement.* Translated by J. H. Bernard. New York: Haefner. Translation of Kant 1790.

Kielmeyer, Carl Friedrich
1793 "Ueber die Verhältnisse der organischen Kräfte unter einander in der Reihe der verschiedenen Organisationen, die Gesetze und Folgen dieser Verhältnisse." Reprinted, with notes by Heinrich Balss, in *Sudhoffs Archiv für Geschichte der Medizin,* 23(1930):247–67.

King, Lester S.
1964 "Stahl and Hoffmann: A Study in Eighteenth-Century Animism." *Journal of the History of Medicine,* 19:118–30.
1969 "Medicine in 1695: Friedrich Hoffmann's *Fundamenta Medicinae.*" *Bulletin of the History of Medicine,* 43:17–29.

Kirchhoff, Alfred
1867 *Die Idee der Pflanzen-metamorphose bei Wolff und bei Göthe.* Berlin: Rudolph Gaertner.
1868 "Caspar Friedrich Wolff. Sein Leben und seine Bedeutung für die Lehre von der organischen Entwickelung." *Jenaische Zeitschrift für Medizin und Naturwissenschaft,* 4:193–220.

Koelreuter, Joseph Gottlieb
1761–66 *Vorläufige Nachricht von einigen das Geschlecht der Pflanzen betreffenden Versuchen und Beobachtungen.* Leipzig: Gleditsch.

Kohlbrugge, J. H. F.
1913 "Historisch-kritische Studien über Goethe als Naturforscher." *Zoologische Annalen*, 5:83–228.
Kubrin, David
1967 "Newton and the Cyclical Cosmos: Providence and the Mechanical Philosophy." *Journal of the History of Ideas*, 28:325–46.
Kuhlemann, Johann Christoph
1753 *Observationes quasdam circa negotium generationis in ovibus factas*. Göttingen: C. L. Schultz.
Kuhn, Thomas S.
1970 *The Structure of Scientific Revolutions*. 2nd ed. Chicago & London: University of Chicago Press.
Lakatos, Imre, and Musgrave, Alan, eds.
1970 *Criticism and the Growth of Knowledge*. Cambridge: Cambridge University Press.
La Mettrie, Julien Offray de
1748 *L'Homme machine*. Leiden: Elie Luzac.
Larson, James L.
1971 *Reason and Experience: The Representation of Natural Order in the Work of Carl von Linné*. Berkeley & London: University of California Press.
1979 "Vital Forces: Regulative Principles or Constitutive Agents? A Strategy in German Physiology, 1786–1802." *Isis*, 70:235–49.
Leibniz, Gottfried Wilhelm von
1951 *Leibniz: Selections*. Edited by Philip P. Wiener. New York: Charles Scribner's Sons.
Lemoine, Albert
1864 *Le Vitalisme et l'animisme de Stahl*. Paris: Baillière.
Lenoir, Timothy
1978 "Generational Factors in the Origin of *Romantische Naturphilosophie*." *Journal of the History of Biology*, 11:57–100.
1980 "Kant, Blumenbach, and Vital Materialism in German Biology." *Isis*, 71:77–108.
Lindeboom, G. A.
1968 *Herman Boerhaave: The Man and His Work*. London: Methuen.
Linnaeus, Carl
1735 *Systema naturae*. Leiden: T. Haak.
1753 *Species plantarum*. Stockholm: L. Salvi. Facsimile reprint ed. London: Ray Society, 1957–59.
Ludovici, Carl Günther
1737 *Ausführlicher Entwurff einer vollständigen Historie der Wolffischen Philosophie*. Leipzig: J. G. Löwe.
Lukina, Tat'jana A.
1975 "Caspar Friedrich Wolff und die Petersburger Akademie der Wissenschaften." *Beiträge zur Geschichte der Naturwissenschaften und der Medizin: Festschrift für Georg Uschmann*, pp. 411–25. Acta Historica Leopoldina, no. 9. Leipzig: J. A. Barth.
Lundsgaard-Hansen-von Fischer, Susanna
1959 *Verzeichnis der gedruckten Schriften Albrecht von Hallers*. Berner Beiträge zur Geschichte der Medizin und der Naturwissenschaften, no. 18. Bern: P. Haupt.

Maître-Jan, Antoine
1722 *Observations sur la formation du poulet*. Paris: L. d'Houry.
Malebranche, Nicolas
1674 *Recherche de la vérité*. Reprinted in *Oeuvres complètes*, vol. 1. Edited by Geneviève Rodis-Lewis. 20 vols. Paris: Librairie Philosophique J. Vrin, 1962.
1688 *Entretiens sur la métaphysique et sur la religion*. Reprinted in *Oeuvres complètes*, vol. 12. Edited by André Robinet. 20 vols. Paris: Librairie Philosophique J. Vrin, 1965.
Malpighi, Marcello
1673 *Dissertatio epistolica de formatione pulli in ovo*. London: J. Martyn. Reprinted and translated in Adelmann 1966, 2: 932–81.
1675 *Anatome plantarum. Cui subjungitur appendix, iteratas et auctas ejusdem authoris de ovo incubato observationes continens*. London: J. Martyn. Appendix reprinted and translated in Adelmann 1966, 2:982–1013.
1686 *Opera omnia*. London: Robert Scott.
Maupertuis, Pierre-Louis Moreau de
1745 *Vénus physique*. n.p. Published anonymously.
1751 *Système de la nature*. In *Oeuvres*, 2:137–68. 4 vols. Lyon: Jean-Marie Bruyset, 1756. Originally published under the pseudonym Dr. Baumann of Erlangen and titled *Dissertatio inauguralis metaphysica de universali naturae systemate*.
1966 *The Earthly Venus*. Translated by Simone Brangier Boas. With notes and an introduction by George Boas. The Sources of Science, no. 29. New York & London: Johnson Reprint Corporation. Translation of Maupertuis 1745.
Mayr, Ernst
1957 "Species Concepts and Definitions." Reprinted in *Evolution and the Diversity of Life*, pp. 493–508. Cambridge, Mass., & London: Harvard University Press, 1976.
1968 "Illiger and the Biological Species Concept." *Journal of the History of Biology*, 1:163–78.
1969 "The Biological Meaning of Species." Reprinted in *Evolution and the Diversity of Life*, pp. 515–25. Cambridge, Mass., & London: Harvard University Press, 1976.
Mazzolini, Renato G.
1976 "Two Letters on Epigenesis from John Turberville Needham to Albrecht von Haller." *Journal of the History of Medicine*, 31:68–77.
1977 "Sugli studi embriologici di Albrecht von Haller negli anni 1755–1758." *Annali dell'Instituto storico italo-germanico in Trento; Jahrbuch des italienisch-deutschen historischen Instituts in Trient*, 3:183–242.
McGuire, J. E.
1968 "Force, Active Principles, and Newton's Invisible Realm." *Ambix*, 15:154-208.
Mendelsohn, Everett
1964 "The Changing Nature of Physiological Explanation in the Seventeenth Century." In *Mélanges Alexandre Koyré*, 1:367–86. 2 vols. Paris: Hermann.

Meyer, Arthur William

1939 *The Rise of Embryology*. Stanford, Calif.: Stanford University Press. London: Humphrey Milford, Oxford University Press.

Moeschlin-Krieg, Beate

1953 *Zur Geschichte der Regenerationsforschung im 18. Jahrhundert*. Basler Veröffentlichungen zur Geschichte der Medizin und der Biologie, no. 1. Basel: Benno Schwabe.

Mulkay, Michael

1979 *Science and the Sociology of Knowledge*. London & Boston: George Allen & Unwin.

Needham, John Turberville

1748 "A Summary of Some Late Observations upon the Generation, Composition, and Decomposition of Animal and Vegetable Substances." *Philosophical Transactions of the Royal Society of London*, 45:615–66.

1750 *Nouvelles Observations microscopiques, avec des découvertes intéressantes sur la composition et la décomposition des corps organisés*. Paris: Louis-Étienne Ganeau.

Needham, Joseph

1934 *A History of Embryology*. Cambridge: Cambridge University Press. Rev. ed. 1959, Cambridge: Cambridge University Press; New York: Abelard-Schuman.

Newton, Isaac

1713 *Philosophiae naturalis principia mathematica*. 2nd. ed. Cambridge: n.p. First edition 1687.

1730 *Opticks*. 4th ed. Reprinted New York: Dover Publications, 1952.

1934 *Sir Isaac Newton's Mathematical Principles of Natural Philosophy and his System of the World*. Translated by Andrew Motte; revised by Florian Cajori. Berkeley & London: University of California Press. Translation of the third edition (1726) of the *Principia*.

Oakley, Francis

1961 "Christian Theology and the Newtonian Science: The Rise of the Concept of the Laws of Nature." *Church History*, 30: 433–57.

Oken, Lorenz

1809–11 *Lehrbuch der Naturphilosophie*. 3 vols. Jena: F. Frommand.

Oppenheimer, Jane M.

1967 *Essays in the History of Embryology and Biology*. Cambridge, Mass., & London: M. I. T. Press.

Ospovat, Dov

1976 "The Influence of Karl Ernst von Baer's Embryology, 1828–1859: A Reappraisal in Light of Richard Owen's and William B. Carpenter's 'Paleontological Application of "Von Baer's Law."'" *Journal of the History of Biology*, 9:1–28.

Pancaldi, Guiliano

1972 *La generazione spontanea nelle prime richerche dello Spallanzani*. Pisa: Domus Galilaeana.

Passmore, J. A.

1958 "William Harvey and the Philosophy of Science." *Australasian Journal of Philosophy*, 36:85–94.

Patten, Bradley M.
1968 *Human Embryology*. 3rd ed. New York: McGraw-Hill.
1971 *Early Embryology of the Chick*. 5th ed. New York: McGraw-Hill.

Pemberton, Henry
1728 *A View of Sir Isaac Newton's Philosophy*. London: S. Palmer.

Perrault, Claude
1680 "La Mechanique des animaux." Reprinted in *Oeuvres diverses de physique et de mechanique*, pp. 329–491. Leiden: Pierre Vander Aa, 1721.

Preus, Anthony
1970 "Science and Philosophy in Aristotle's *Generation of Animals*." *Journal of the History of Biology*, 3:1–52.
1977 "Galen's Criticism of Aristotle's Conception Theory." *Journal of the History of Biology*, 10:65–85.

Price, Lawrence Marsden
1926 "Albrecht von Haller and English Theology." *Publications of the Modern Language Association*, 41:942–54.

Procès-verbaux des séances de l'académie impériale des sciences depuis sa fondation jusqu'à 1803. Protkoly zasedanij konferencij imperatorskoj akademij nauk s 1725 po 1803 goda.
1897–1911 4 vols. St. Petersburg: n.p.

Punnett, R. C.
1928 "Ovists and Animalculists." *American Naturalist*, 62:481–507.

Rajkov, B. E.
1964 "Caspar Friedrich Wolff." *Zoologische Jahrbücher*, 91:555–626. Translation of a chapter from *Ocherki po istorii evoljucionnoj idei v Rossii do Darvina* [Survey of a history of evolutionary ideas in Russia up to Darwin]. Moscow & Leningrad: Akademii nauk SSRR, 1947.

Rather, L. J.
1961 "G. E. Stahl's Psychological Physiology." *Bulletin of the History of Medicine*, 35:37–49.

Ray, John
1686–1704 *Historia plantarum*. 3 vols. London: M. Clarke.

Réaumur, René Antoine Ferchault de
1712 "Sur les diverses reproductions qui se font dans les écrevisses, les omars, les crabes, etc. et entr' autres sur celles de leurs jambes et de leurs écailles." *Mémoires de l'Académie Royale des Sciences*, pp. 223–41. Paris: L'Imprimerie Royale.
1734–42 *Mémoires pour servir à l'histoire des insectes*. 6 vols. Paris: L'Imprimerie Royale.
1749 *Art de faire éclorre et d'élever en toute saison des oiseaux domestique de toutes espèces*. 2 vols. Paris: L'Imprimerie Royale.
1750 *The Art of Hatching and Bringing up Domestic Fowls of all Kinds, at any Time of the Year*. Translated by Abraham Trembley. London: C. Davis, A. Millar, and J. Nourse. Translation of Réaumur 1749.

Richter, Karl
1972 *Literatur und Naturwissenschaft: Eine Studie zur Lyrik der Aufklärung*. Munich: Wilhelm Fink.

Roe, Shirley A.

1975 "The Development of Albrecht von Haller's Views on Embryology." *Journal of the History of Biology*, 8:167–90.

1979 "Rationalism and Embryology: Caspar Friedrich Wolff's Theory of Epigenesis." *Journal of the History of Biology*, 12:1–43.

1981 "*Anatomia animata*: The Newtonian Physiology of Albrecht von Haller." In *Transformation and Tradition in the Sciences*. Edited by Everett Mendelsohn. Forthcoming.

Roger, Jacques

1963 *Les Sciences de la vie dans la pensée française du XVIIIe siècle: La génération des animaux de Descartes à l'Encyclopédie*. Paris: Armand Colin. 2nd ed., 1971.

1968 "Leibniz et les sciences de la vie." *Studia Leibnitiana Supplementa*, 2:209–19.

1973 "Buffon, Georges-Louis Leclerc, Comte de." *Dictionary of Scientific Biography*, 2:576–82. New York: Charles Scribner's Sons.

Rostand, Jean

1930 *La Formation de l'être: Histoire des idées sur la génération*. Paris: Librairie Hachette.

Rouhault, Pierre Simon

1728 *Réponse de Pierre Simon Rouhault à la critique faite à son mémoire de la circulation du sang dans le foetus humain, par M. Winslow*. Turin: J. F. Mairesse.

Rudolph, Gerhard

1964 "Hallers Lehre von der Irritabilität und Sensibilität." In *Von Boerhaave bis Berger*, pp. 14–34. Edited by K. E. Rothschuh. Medizin in Geschichte und Kultur, vol. 5. Stuttgart: Gustav Fischer.

Saussure, Raymond de

1949 "Haller and La Mettrie." *Journal of the History of Medicine*, 4:431–49.

Savioz, Raymond

1948a *Mémoires autobiographiques de Charles Bonnet de Genève*. Paris: Librairie Philosophique J. Vrin.

1948b *La Philosophie de Charles Bonnet de Genève*. Paris: Librairie Philosophique J. Vrin.

Schär, Rita

1958 *Albrecht von Hallers neue anatomisch-physiologische Befunde und ihre heutige Gültigkeit*. Berner Beiträge zur Geschichte der Medizin und der Naturwissenschaften, no. 16. Bern: P. Haupt.

Scheffler, Israel

1967 *Science and Subjectivity*. New York: Bobbs-Merrill.

Schopfer, W. H.

1945 "L'Histoire des théories relatives à la génération, aux 18ème et 19ème siècles." *Gesnerus*, 2:81–103.

Schrecker, Paul

1938 "Malebranche et le préformisme biologique." *Revue internationales de philosophie*, 1:77–97.

Schuster, Julius

1936 "Caspar Friedrich Wolff. Leben und Gestalt eines deutschen Biologen." *Sitzungsberichte der Gesellschaft Naturforschender Freunde zu Berlin*, pp. 175–95.

1941 "Der Streit um die Erkenntnis des organischen Werdens im Lichte der Briefe C. F. Wolffs an A. von Haller." *Sudhoffs Archiv für Geschichte der Medizin*, 34:196–218.

Schütz, V.

1947 "Kaspar Friedrich Wolff in Russland." *Experientia*, 3:465–67.

Sigerist, Henry W., ed.

1923 *Albrecht von Hallers Briefe an Johannes Gessner (1728–1777)*. Abhandlung der Königlichen Gesellschaft der Wissenschaften zu Göttingen, Math.-Physik. Klasse, n.s., 11, no. 2.

Sloan, Philip R.

1977 "Descartes, the Sceptics, and the Rejection of Vitalism in Seventeenth-Century Physiology." *Studies in History and Philosophy of Science*, 8:1–28.

Sonntag, Otto

1971 "The Idea of Natural Science in the Thought of Albrecht von Haller." Ph.D. dissertation, New York University.

1974a "Albrecht von Haller on the Future of Science." *Journal of the History of Ideas*, 35:313–22.

1974b "The Motivations of the Scientist: The Self-Image of Albrecht von Haller." *Isis*, 65:336–51.

1975 "Albrecht von Haller on Academies and the Advancement of Science: the Case of Göttingen." *Annals of Science*, 32:379–91.

1977 "The Mental and Temperamental Qualities of Haller's Scientist." *Physis*, 19:173–84.

Spallanzani, Lazzaro

1765 *Dissertazioni due*. Contains *Saggio di osservazioni microscopiche concernenti il sistema della generazione dei Signori di Needham e Buffon*. Modena: B. Soliani.

1769 *Nouvelles Récherches sur les découvertes microscopiques, et la génération des corps organisés*. Translated by Abbé Regley, with a critical commentary by John Turberville Needham. London & Paris: Lecombe. Translation of Spallanzani 1765.

1776 *Opuscoli di fisica animale e vegetabile*. 2 vols. Modena: Società Tipografica.

Stahl, Georg Ernst

1708 *Theoria medica vera*. Halle: Literis Orphanotrophei.

Stearn, W. H.

1957 Introduction to *Species Plantarum* (1753) by Carl Linnaeus. Facsimilie reprint ed. London: Ray Society.

Stresemann, E.

1962 "Leben und Werk von Peter Simon Pallas." In *Lomonosov, Schlözer, Pallas: Deutsch-Russische Wissenschaftsbeziehungen im 18. Jahrhundert*, pp. 247–57. Edited by Eduard Winter. Berlin: Academie-Verlag.

Sturm, Friedrich August Bernhard

1974 *Albrecht von Hallers Lehre über die Enstehung der Missbildungen*. Bonn: n.p.

Suppe, Frederick, ed.
1974 *The Structure of Scientific Theories*. Introduction by Frederick Suppe. Urbana: University of Illinois Press.

Swammerdam, Jan
1669 *Historia insectorum generalis*. Utrecht: Van Dreunen.
1672 *Miraculum naturae, sive uteri muliebris fabrica*. Leiden: S. Matthaei.

Tamney, Martin
1979 "Newton, Creation, and Perception." *Isis*, 70:48–58.

Teeter, Lura May
1928 "Albrecht von Haller and Samuel Clarke." *Journal of English and Germanic Philology*, 27:520–23.

Temkin, Owsei
1950 "German Concepts of Ontogeny and History Around 1800." *Bulletin of the History of Medicine*, 24:227–46.
1963 "Basic Science, Medicine, and the Romantic Era." *Bulletin of the History of Medicine*, 37:97–129.

Toellner, Richard
1967 "Anima et Irritabilitas, Hallers Abwehr von Animismus und Materialismus." *Sudhoffs Archiv für Geschichte der Medizin*, 51:130–44.
1971 *Albrecht von Haller: Über die Einheit im Denken des letzten Universalgelehrten*. Sudhoffs Archiv Beihefte, no. 10. Wiesbaden: F. Steiner.
1972 "The Controversy between Descartes and Harvey regarding the Nature of Cardiac Motion." In *Science, Medicine, and Society in the Renaissance: Essays to Honor Walter Pagel*, 2:73–89. Edited by Allen G. Debus. 2 vols. New York: Science History Publications.
1973 "Haller und Leibniz, zwei Universalgelehrte der Aufklärung." *Studia Leibnitiana Supplementa*, 12:249–60.
1976 "Staatsidee, aufgeklärter Absolutismus und Wissenschaft bei Albrecht von Haller." *Medizinhistorisches Journal*, 11:206–19.

Tournefort, Joseph Pitton de
1694 *Élémens de botanique ou méthode pour connoître les plantes*. 3 vols. Paris: L'Imprimerie Royale.

Trembley, Abraham
1744 *Mémoires pour servir à l'histoire d'un genre de polypes d'eau douce, à bras en forme de cornes*. Leiden: Verbeek.

Trembley, Maurice, ed.
1943 *Correspondance inédite entre Réaumur et Abraham Trembley*. With an introduction by Émile Guyénot. Geneva: Georg.

Trew, Christoph Jacob
1736 *Dissertatio epistolica de differentiis quibusdam inter hominem natum et nascendum intercedentibus deque vestigiis divini numinis inde colligendis*. Nuremberg: P. C. Monath.

Uschmann, Georg
1955 *Caspar Friedrich Wolff: Ein Pioneer der modernen Embryologie*. Leipzig & Jena: Urania.
1962 "C. F. Wolff und Pallas." In *Lomonosov, Schlözer, Pallas: Deutsch-Russische Wissenschaftsbeziehungen im 18. Jahrhundert*,

pp. 315–17. Edited by Eduard Winter. Berlin: Acadmie-Verlag.

Vartanian, Aram
1949 "Elie Luzac's Refutation of La Mettrie." *Modern Language Notes*, 64:159–61.
1950 "Trembley's Polyp, La Mettrie, and Eighteenth-Century French Materialism." *Journal of the History of Ideas*, 11:259–86.
1953 Review of *Abraham Trembley of Geneva: Scientist and Philosopher* by John R. Baker. *Isis*, 44:387–89.

Webster, C.
1966–67 "Harvey's *De Generatione*: Its Origins and Relevance to the Theory of Circulation." *British Journal of the History of Science*, 3:262-74.

Weismann, August
1891–92 "The Supposed Transmission of Mutilations." In *Essays upon Heredity and Kindred Biological Problems*, 1:431–61. Translated and edited by Edward B. Poulton, Selmer Schönland, and Arthur E. Shipley. 2 vols. 2nd. ed. Oxford: Clarendon Press.

Wheeler, William Morton
1899 "Caspar Friedrich Wolff and the *Theoria Generationis*." *Biological Lectures from the Marine Biological Laboratory, Wood's Hole, Mass., 1898*, pp. 265–84. Boston: Ginn.

Whitman, C. O.
1895a "Bonnet's Theory of Evolution: A System of Negations." *Biological Lectures Delivered at the Marine Biological Laboratory of Wood's Hole, 1894*, pp. 225–40. Boston: Ginn.
1895b "The Palingenesia and the Germ Doctrine of Bonnet." *Biologial Lectures Delivered at the Marine Biological Laboratory of Wood's Hole, 1894*, pp. 241–72. Boston: Ginn.

Whytt, Robert
1751 *An Essay on the Vital and Other Involuntary Motions of Animals*. Edinburgh: Hamilton, Balfour, & Neill. 2nd ed. 1763.
1755 *Physiological Essays*. Edinburgh: Hamilton, Balfour, & Neill. 2nd ed. 1761.

Wilkie, J. S.
1967 "Preformation and Epigenesis: A New Historical Treatment." *History of Science*, 6:138–50.

Wilson, J. Walter
1944 "Cellular Tissue and the Dawn of the Cell Theory." *Isis*, 35:168–73.

Winslow, Jacob
1717 "Description d'une valvule singuliere de la veine-cave inferieur." *Mémoires de l'Académie Royale des Sciences*, pp. 211-25. Paris: L'Imprimerie Royale. Published 1719.
1725a "Éclaircissemens sur un mémoire de 1717, qui traite de la circulation du sang dans le foetus." *Mémoires de l'Académie Royale des Sciences*, pp. 23–34. Paris: L'Imprimerie Royale. Published 1727.
1725b "Suite des éclaircissemens sur la circulation du sang dans le foetus." *Mémoires de l'Académie Royale des Sciences*, pp. 260–81. Paris: L'Imprimerie Royale. Published 1727.

Wolff, Caspar Friedrich
1759 *Theoria generationis*. Halle: Hendel. Facsimile reprint ed. (with Wolff 1764), Hildesheim: Georg Olms, 1966.
1764 *Theorie von der Generation in zwo Abhandlungen erklärt und bewiesen*. Berlin: Friedrich Wilhelm Birnstiel. Fascimile reprint ed. (with Wolff 1759), Hildesheim: Georg Olms, 1966.
1766–67 "De formatione intestinorum praecipue, tum et de amnio spurio, aliisque partibus embryonis gallinacei, nondum visis, observationes, in ovis incubatis institutae." Parts 1 and 2. *Novi Commentarii Academiae Scientiarum Imperialis Petropolitanae*, 12:403–507. Published 1768.
1768 "De formatione intestinorum. Observationes in ovis incubatis institutae." Part 3. *Novi Commentarii Academiae Scientiarum Imperialis Petropolitanae*, 13:478–530. Published 1769.
1772 "Descriptio vituli bicipitis cui accedit commentatio de ortu monstrorum." *Novi Commentarii Academiae Scientiarum Imperialis Petropolitanae*, 17:540–75. Published 1773.
1774 *Theoria generationis*. 2nd. ed. Halle: Hendel.
1775 "De foramine ovali, eiusque usu, in dirigendo motu sanguinis. Observationes novae." *Novi Commentarii Academiae Scientiarum Imperialis Petropolitanae*, 20:357–430. Published 1776.
1778 "Notice touchant un monstre biforme, dont les deux corps sont réunis par derriere." *Acta Academiae Scientiarum Imperialis Petropolitanae*, 2, pt. 1:41–44. Published 1780.
1780 "De pullo monstroso, quatuor pedibus, totidemque alis instructo." *Acta Academiae Scientiarum Imperialis Petropolitanae*, 4, pt. 1:203–7. Published 1783.
1789 *Von der eigenthümlichen und wesentlichen Kraft der vegetabilischen sowohl als auch der animalischen Substanz*. St. Petersburg: Kayserliche Academie der Wissenschaften. Published with Blumenbach and Born 1789.
1812 *Über die Bildung des Darmkanals im bebrüteten Hünchen*. Translated and with an introduction by Johann Friedrich Meckel. Halle: Renger. Translation of Wolff 1766–67, 1768.
1896 *Caspar Friedrich Wolff's "Theoria Generationis" (1759)*. Translated by Paul Samassa. Ostwalds Klassiker, nos. 84 and 85. Leipzig: Wilhelm Engelmann. German translation of Wolff 1759.
1950 *Theorija zarozhdenija*. Edited by E. N. Pavlovskij, with notes and commentary by A. E. Gaissinovitch. Moscow: Izdatel'stvo akademii nauk SSSR. Translation of Wolff 1759 and portions of Wolff 1764.
1973 *Objecta meditationum pro theoria monstrorum; Predmety razmyshlenij v svjazi s teoriej urodov*. Translated by Ju. Kh. Kopelevich and T. A. Lukina. With notes by T. A. Lukina. Leningrad: Izdatel'stvo ≪Nauka≫.

Wolff, Christian
1723 *Vernünfftige Gedancken von der Würckungen der Natur*. Halle: Renger.
1728 *Philosophia rationalis sive logica, methodo scientifica pertractata et ad usum scientiarum atque vitae aptata*. Frankfurt & Leipzig: Renger.

1963 *Preliminary Discourse on Philosophy in General.* Translated and with an introduction and notes by Richard J. Blackwell. Library of Liberal Arts. Indianapolis & New York: Bobbs-Merrill. Translation of "Discursus praeliminaris" in Christian Wolff 1728.

Zimmermann, Johann Georg

1755 *Das Leben des Herrn von Haller.* Zürich: Heidegger.

Index

www.ingramcontent.com/pod-product-compliance
Ingram Content Group UK Ltd.
Pitfield, Milton Keynes, MK11 3LW, UK
UKHW040705180125
453697UK00010B/419